Virtual Reality in Higher Education

Virtual Reality in Higher Education

Instruction for the Digital Age

Edited by

Darrel W. Staat

AMERICAN ASSOCIATION OF COMMUNITY COLLEGES

ROWMAN & LITTLEFIELD

Lanham • Boulder • New York • London

Published by Rowman & Littlefield
An imprint of The Rowman & Littlefield Publishing Group, Inc.
4501 Forbes Boulevard, Suite 200, Lanham, Maryland 20706
www.rowman.com

6 Tinworth Street, London, SE11 5AL, United Kingdom

British Library Cataloguing in Publication Information Available

Library of Congress Cataloging-in-Publication Data

Names: Staat, Darrel W., 1941– editor.
Title: Virtual reality in higher education: instruction for the digital age / Edited by Darrel W. Staat.
Description: Lanham: Rowman & Littlefield, [2021] | Includes bibliographical references. | Summary: "Virtual Reality in Higher Education: Instruction for the Digital Age contains eight chapters of graduate student research about how virtual reality is being used in institutions of higher education in specific areas of education, training, and athletic recruitment."—Provided by publisher.
Identifiers: LCCN 2021009162 (print) | LCCN 2021009163 (ebook) | ISBN 9781475861280 (cloth) | ISBN 9781475861297 (paperback) | ISBN 9781475861303 (epub)
Subjects: LCSH: Virtual reality in higher education. | Computer-assisted instruction. | College athletes—Recruiting.
Classification: LCC LB2395.7.S77 2021 (print) | LCC LB2395.7 (ebook) | DDC 378.1/7344678—dc23
LC record available at https://lccn.loc.gov/2021009162
LC ebook record available at https://lccn.loc.gov/2021009163

Contents

Preface

As a university professor, I am always on the lookout for better ways of teaching the graduate students who enroll in my classes. I still use the tried and true lecture/discussion method to some extent, but over the years I have added the flipped classroom to my repertoire with considerable success for the students. When I was first introduced to the teaching potential of virtual reality (VR) I was literally astounded. The head-mounted displays (HMD) on the market today offer possibilities for immersive learning that were not possible just a few years ago.

In my initial research I found that K–12 schools were already experimenting with it during the past few years. Inexpensive HMDs such as the Google Cardboard allow students to enter the VR world in a variety of subject matters. My research into the use of VR in higher education was spottier. VR did exist in some institutions, but at this point in time is viewed with a modicum of suspicion as just another new technology that will come and go with little impact.

The graduate students in my course Higher Education in the 21st Century have been researching institutions of higher education to determine which are already experimenting with VR as a teaching method and what kinds of success they are having. Their findings were quite significant, as described in this book. My sincere hope it that faculty in higher education will become aware of the potential of VR for instruction and begin using it in appropriate settings. The possibilities are vast and powerful. Virtual reality has the potential to significantly change education as we know it.

Acknowledgments

I would like to thank Dr. Travis Teague, dean and vice provost at Wingate University, for his interest in and support of my work in developing this book.

I would also thank the graduate students in Cohort 6 who have researched and written eight of the chapters in this book. They became intrigued with the subject and really dug into the topic with panache. I appreciate their efforts very much.

My thanks to my colleagues in the Graduate Education Department who have encouraged me. I am in their debt.

My wife Beverly has been a solid support through the long hours of research, writing, and editing of this book. Her patience and understanding are greatly appreciated.

Introduction

Merriam-Webster defines virtual reality as "an artificial environment that is experienced through sensory stimuli (as sights and sounds) provided by an interactive computer program" (Merriam-Webster, 2004, p. 809). Virtual reality (VR) has been around in various forms for almost 200 years. Historically, its beginnings go back to the 1830s, when "inventors began tinkering with stereoscopes that used optics and mirrors along with a pair of lenses to produce a 3D view of objects" (Greengard, 2019, p. 2).

First produced in 1939, the View-Master, a handheld stereoscopic viewer, became very popular: "in 1961 the world's first head-mounted display (HMD) appeared" (Greengard, 2019, p. 16). In 2010, Oculus Rift was created. Today there is range of head-mounted displays from Google Cardboard, a very inexpensive HMD that is used in public education in many K–12 schools, to Oculus Rift, HTC, and Halolens, expensive high-end HMDs.

The military has found the development of VR useful for training its recruits. The aircraft industry has used VR training since the 1930s with the invention of the Link trainer and the many improvements to that process over the decades. Today, as this book will demonstrate, VR is already being used in a variety of ways for instruction. However, higher education seems to be only at the beginning of interest in this powerful teaching method.

The purpose of this book is to enlighten educators in community colleges and universities of the amazing potential VR has as a teaching method. Students using the method become totally engrossed in an immersive experience that parallels reality itself. VR has the ability to stand the usual teaching method of lecture/discussion in higher education on its head. It will allow the student to go from hearing or reading about a subject to becoming immersed by experiencing the topic as if it were actually happening. The impression on the mind and memory will be many times that of a written, verbal, or video method.

VR is a teaching method that faculty in higher education must become acquainted with and begin to experimentally use in their teaching. Higher education today is standing on the brink of a very valuable teaching process that has the potential to make over instruction with phenomenal results for the student, the faculty member, and learning itself. Do not allow this opportunity to be left on the table. Faculty in almost any field should give it a try to see just how powerful it is. With immersive VR, seeing, hearing, and feeling are believing.

REFERENCES

Greengard, S. (2019). *Virtual Reality*. Cambridge, MA: The MIT Press.

Merriam-Webster's Collegiate Dictionary, eleventh edition. (2004). Springfield, MA: Merriam-Webster, Inc.

Chapter 1

Nursing

Tou Vang

Nursing education is no stranger to emerging technologies it has used for teaching in the classroom. For years, nursing programs have been at the forefront of adopting and using healthcare simulation to provide a safe learning environment for nursing students before they enter the workforce. The benefits of implementing these technologies have assisted faculty immensely by allowing students to learn by doing. As the number of clinical sites continues to diminish, simulation-based learning is rapidly becoming a more popular alternative for many nursing programs.

Virtual reality (VR) is one of many technologies that hope to bridge the education gap between knowledge and application for nursing students. Nursing programs across the country are already embracing this technology and have incorporated it as part of their curriculum. It is quickly becoming the next innovation for nursing education.

HISTORY OF SIMULATION BEFORE VIRTUAL REALITY

Simulation in healthcare has a long history. In 1911, the very first life-sized cardiopulmonary resuscitation (CPR) simulation mannequin was created for training healthcare workers; it was known as "Mrs. Chase" (Weir, 2012). Later in the 1960s, Resusci Anne was introduced and became one of the most successful and iconic CPR training mannequin for healthcare professionals, even to this day (Cooper & Taqueti, 2008).

In 1966, the first computer-controlled patient simulator was created. Sim One was developed by Dr. Stephen Abrahamson and Dr. Judson Denson at the University of Southern California. However, it was not widely accepted and was too expensive to commercialize. It was not until the mid-1990s that the SimMan Classic was developed. It was one of the first, substantially less expensive, high-fidelity mannequin developed by Dr. Rene Gonzales and Dr. John Schaefer, in partnership with the University of Pittsburgh and Medical Plastics Corporation (MPL), which Laerdal later acquired (Cooper & Taqueti, 2008).

Since then, manufacturers such as CAE Healthcare, Gaumard, and Laerdal have started developing a multitude of computerized commercial mannequins, many of which are aimed at nursing education. Some of the most notable are iStan by CAE, SimMan 3G by Laerdal, and Noelle and HAL by Gaumard. Many of these are referred to as high-fidelity mannequins, which are full-body mannequins that can mimic, at a very high level, human body functions (Lioce et al., 2020).

IMPORTANCE OF SIMULATION-BASED LEARNING

Before institutions start implementing VR into nursing education, they first need to understand the importance of simulation-based learning. Similar to other programs, nursing has strict guidelines on how simulation can be used in the clinical lab or curriculum courses.

There are many definitions of simulation-based learning used by both healthcare professionals and the tech industry. The most commonly used for healthcare is "an array of structured activities that represent actual or potential situations in education and practice and allows participants to develop or enhance knowledge, skills, and attitudes or to analyze and respond to realistic situations in a simulated environment" (Lioce et al., 2020). This definition summarizes the approach that many nursing programs have moved to, which is the use of simulators to enhance learning.

Patient simulators are common in nursing education today and have been for decades. Many of these high-fidelity mannequins are equipped with realistic physiologic responses including but not limited to respiration, pulses, breath sounds, heart sounds, urinary output, pupil reaction, and ability to communicate to students. They are controlled using a personal computer (PC), which allows for adjustments to the patient scenarios to be made on the fly by the nurse educator (Durham & Alden, 2008).

Some of the advantages of using patient simulation in nursing education are:

- Immersive, experiential learning that provides an opportunity for active and interactive learning without risk to an actual patient

- Preventing medication errors due to inappropriate medication dosage, overlooked allergies, giving the wrong drug, or incorrect administration site
- Developing critical thinking and clinical decision-making skills to assess, plan, implement, and evaluate nursing care
- Promoting effective communication among caregivers during times of patient transition in healthcare settings
- Encouraging teamwork to delegate tasks appropriately, to follow directions, and to communicate effectively with nurses and other practitioners
- Boosting students' self-confidence and reducing anxiety in an actual patient care setting (Durham & Alden, 2008, p. 223)

Why are so many nursing schools using simulation-based learning? One of the reasons is the decrease in the availability of clinical training sites. According to the American Association of Critical Care Nursing (AACN)'s report on 2016–2017, Enrollment and Graduations in Baccalaureate and Graduate Programs in Nursing, U.S. nursing schools turned away 64,067 qualified applicants from baccalaureate and graduate nursing programs in 2016 due to insufficient number of faculty, clinical sites, classroom space, and clinical preceptors, as well as budget constraints (AACN Fact Sheet: Nursing Shortage, 2017).

Almost two-thirds of the nursing schools responding to the survey pointed to a shortage of faculty and/or clinical preceptors as a reason for not accepting all qualified applicants into their programs (AACN Fact Sheet: Nursing Shortage, 2017). Simulation-based learning offers a unique approach to tackling the shortage of clinical training sites that many nursing schools are struggling to secure. If fact, many nursing schools have already replaced some clinical hours with simulation.

Of the 50 states, 13 boards of nursing (BON) have allowed up to 50% of clinical hours to be replaced by simulation (FL, IA, KY, LA, MN, NH, NM, SC, SD, TN, TX, WA, and WI). Other state BONs allow smaller percentages such as 30% (DC and OK) and 25% (CA, IL, IN, MS, NV, VT, and VA). The remaining states either did not have a percentage range or did not specify. For example, North Carolina's BON only allows 25% for focused client care and no more than 50% in other clinical experience; all other simulation is limited to 25% if the program is not accredited (Bradley et al., 2019).

Simulation is accelerating across many nursing programs with over 100 accredited simulation centers in the United States and growing. These sim centers are accredited by the Society for Simulation in Healthcare (SSH), the largest healthcare simulation accrediting body in the world (SSH, 2020).

In addition, nurse educators who are committed to the continuous improvement of healthcare simulation can also receive professional recognition by applying to be a certified healthcare simulation educator (CHSE), certified healthcare simulation operations specialist (CHSOS), or a certified healthcare simulation educator-advanced (CHSE-A) (SSH, 2020).

Without a doubt, simulation-based learning has disrupted the way nursing programs have integrated simulation into their curriculum and clinical labs for decades. With the introduction of high-fidelity mannequins, standards of best practice for simulation were created by the International Nursing Association for Clinical Simulation and Learning (INACSL) to ensure best practices in design, implementation, and evaluation of simulation activities. However, nursing programs most often still look to the BONs for clear regulatory guidance on the use of simulation as a form of clinical education (Bradley et al., 2019).

The question is, can virtual reality have the same effect? Can it offer a high level of simulation experience, similar to high-fidelity mannequins? Can it change the way nursing students learn in the classroom? And will it meet the BON's regulation and guidelines for substituting for traditional clinical learning hours?

WHAT IS VR IN NURSING?

Just like simulation-based learning, there are many ways to define virtual reality. According to the Society for Simulation in Healthcare, virtual reality is a computer-generated three-dimensional environment that gives an immersion effect (Lioce et al., 2020). This means that in order for a technology to qualify as VR, it must meet two main criteria in healthcare simulation. First, it must create a three-dimensional environment. And second, it must have an immersion effect.

This is crucially important to distinguish between technologies that are marketing themselves as virtual, which may confuse many nurse educators and give the impression that it is VR. Three-dimensional environment is defined as being lifelike such as having length, width, and depth (Merriam-Webster, n.d.). Immersion is being able to place a human in a synthetic environment through physical and/or emotional methods (Modeling and Simulation Glossary, 2016). This means that a physical interface is needed, such as a head-mounted display or haptic devices, to be classified as VR.

For example, vSim for Nursing, codeveloped by Laerdal and Wolters Kluwer Health, is a very popular online simulation program used by many nursing programs. It is sold as part of Lippincott CoursePoint+, an online curriculum platform, and is available as a stand-alone product. Most nursing programs that use Lippincott Williams & Wilkins textbooks and products, now owned by Wolters Kluwer, will receive this add-on access to vSim (Wolters

Kluwer Health, 2014). The program includes virtual simulation patients, each with different scenarios that allow nursing students to access and practice patient care anytime and anywhere.

Students are given various patient care decisions to choose from, and the avatar will react based on what action was chosen (A. Hamby, personal communication, June 5, 2019). Because the word "virtual" is included in the program name, there are assumptions that vSim is a form of VR. But it is not. Even though a three-dimensional environment is created, it does not physically immerse students. Other programs similar to vSim are Shadow Health and vClinical. Both are widely used in nursing education (A. Hamby, personal communication, June 5, 2019).

Though these programs are highly valued and used by many nursing programs, it is essential to understand that they are not a form of VR. Additionally, this includes augmented reality (AR) such as the Microsoft HoloLens. AR is a form of technology that overlays digital computer-generated information on objects or places in the real world for the purpose of enhancing the user experience. It is a combination of reality and digital information only and should not be considered as a VR technology (Lioce et al., 2020). Pokémon Go, the mobile gaming app, is a perfect example of AR.

CURRENT VR TECHNOLOGY

This section will outline some of the technologies that are currently available and being used in nursing education. These technologies can help enhance learning in the classroom and also replace clinical hours for nursing students.

360-degree camera. There are various 360-degree video-capturing devices today. Many of them are portable and will allow users to capture 360-degree images or videos, which provides individuals with a view as if they are physically there. It has mostly been used in real estate for virtual home or apartment tours. A significant advantage of a 360-degree camera is the ability to integrate VR into it. With the correct software and hardware, it can support VR and allow users to use whatever device they have available, whether that is a headset or smartphone.

Oculus Rift. Similar to the HTC, the Oculus Rift from Facebook is a commercially available VR headset with the same capabilities to immerse users in a 3D world. It is a great technology that is already commercially available. Rather than develop the hardware, educators can focus on just the software instead and use the Oculus Rift's advanced technology to power it. The headset also comes with controllers but with a bonus of integrated earphones at a lower starting price compared to the HTC.

VR goggles. There are various types of goggles or headsets that can fully immerse individuals in a VR world, many with just their smartphone. Most of these headsets can be bought online or locally for a relatively small price. For example, Google cardboard is a popular option for many. A simple search on the Internet offers many choices to choose from. This is an excellent option for applications that can take full advantage of smartphone technologies without the need for a high-end headset.

CURRENT USAGE OF VR IN NURSING EDUCATION

Many schools have jumped on the bandwagon of VR for the limitless possibilities that it can bring to the classroom and in clinical labs. It is up to nursing programs to decide whether or not VR can offer the same benefits as simulation-based learning with patient simulators. Below are some examples of how VR is currently being used in nursing classrooms and clinical labs.

Indiana University-Purdue University Columbus School of Nursing

The Indiana University-Purdue University Columbus (IUPUC) simulation center was one of the first few institutions to introduce VR technology into clinical labs for nursing school in 2018. IUPUC identified the technology as an affordable, efficient replacement for many of the expensive mannequins and quickly added it to their lab. Now, their nursing students can use VR to practice clinical scenarios in a safe learning space while providing an invaluable experience (Bratton, 2019).

IUPUC's School of Nursing worked with SimX to custom design scenarios to fit into their current nurse courses to provide experience in both clinical and personal situations. The VR headset, Sixsense mobile VR, includes hand controllers, base stations, and head tracking to be used with a smartphone (Chau, 2017). This means that the students can experience the VR wirelessly and move around the room freely with no additional wires connected to a PC.

One scenario developed for the lab immerses nursing students into a neonatal abstinence syndrome case. This is a common situation in the region where an infant is born to a mother who is addicted to drugs. As students work through the scenario and ask questions, the instructor will choose patient responses from a programmed list to respond to the students through the headset. The purpose of this scenario is to focus on how the student communicates with the mother and is less focused on the baby (Bratton, 2019).

The VR technology has certainly provided real-life situations to second-year nursing students, and they appreciate the reality aspect. They are communicating with not only a patient but also family members, which typically happens in real life. The technology has allowed faculty to place it in the curriculum and cater to the needs of IUPUC nursing students. A limitation of clinical sites is that, if students are not in the right place at the right time, they will not learn anything. However, with VR, the technology can bring the right time to the students (Bratton, 2019).

The Valley Foundation School of Nursing at San Jose State University

Instead of using high-end devices like the HTC or Oculus Rift, some institution of higher education have opted to go with a lower-end device, Syminar, to make it more accessible to students. Syminar is a live 360-degree video capture and broadcasting service. Syminar allows nurse educators to capture and stream sessions to students through the program's online platform using a 360-degree camera. The live video can be assessed on any device or in VR mode with the included VR goggles, which uses a smartphone. Sessions are recorded and can be used for debriefing or study review afterward (Syminar, 2018).

Syminar has been used to practice clinical situations in labs for some time now. Students can come onto campus and use the VR in the debriefing room or access it from their phone with the VR goggles. During the session, students can look around and view all the equipment available to them for the specific scenario such as in a medical-surgical (med-surg), obstetrics (OB), or pediatric setting. A technician does all the moving for the students by rolling the 360-degree video camera around (D. Nelson, personal communication, June 19, 2019).

Before each session, the technician would set up the course and develop the pre-briefing, charts, and debriefing materials for students. Post-simulation clinical-thinking questions are also conducted either in person or through live debrief using Zoom. The only downside to Syminar is that students cannot interact in the actual session when using the VR goggles. They can only type questions or answers. Voice is currently not supported (D. Nelson, personal communication, June 19, 2019).

Syminar's use of the 360-degree camera fully immerses students in the lab and allows them to look around the room as if they were physically there. As they look around, students can decide what they can do to meet objectives during the session (D. Nelson, personal communication, June 5, 2019). This is a great approach to using the 360-degree camera technology, as it is more affordable and accessible to students.

Penn State

Penn State is one of few institutions to incorporate VR into its online classes to help nursing students learn more effectively. Designers worked with professors in the College of Nursing to incorporate a 360-degree video into Nursing 352, a course on Advanced Health Assessment, which is offered online through the Penn State World Campus RN to BSN nursing program. Students in the class were offered a free VR headset to use with their smartphone. As part of the course, students learned to assess a patient's home environment for safety, cleanliness, and other factors that should be addressed with their patients (Appelman, 2018).

One scenario shows students the home of an elderly woman from the viewpoint of a nurse who is making the home visit. Students can explore the different rooms such as the bathroom, kitchen, and living room, checking for hazards or improperly stored medications. Students can use the VR headset with their smartphone and fully immerse themselves as if they were in the room. This experience allowed students to practice skills as if they were in a real-life situation, which helps them learn more effectively and efficiently (Appelman, 2018).

Researchers will continue to collect feedback and data from students in all the Penn State World Campus courses that are using the VR. If the 360-degree video proves to be successful, designers will move forward with developing true VR content where users can interact with the surroundings (Appelman, 2018).

Boise State Universmity

Boise State University has developed a game to help give nursing students practice in performing common medical procedures. The technology is called Virtual Reality Nursing Simulation with Custom Haptic System for Patient Safety. It uses the Oculus Rift with custom haptic gloves to recreate the sense of touch for users. The simulation walks students through a VR environment and then scores them on how well they complete a series of tasks, including sterilizing their environment (Madrid, 2016).

The main goal of the game is to allow nursing students to practice and properly insert a sterile urinary catheter. Catheterization is a good candidate for VR training because it is difficult and is dangerous to patients if not done correctly. In the VR scenario, students must go to one corner of the room to get a tray of the sterile package. Then they need to wash their hands, put on clean gloves, pick up the catheter package, and bring it back to the patient. After cleaning the patient's bottom, they need to properly open the catheter package and insert the sterile catheter correctly. A green cloud of small, falling particles will appear on the screen if students did not wear their gloves properly (He, 2018).

Nursing students can practice this skill as many times as needed until they reach a desired level of proficiency. Since its initial release, the game has developed into a second version, which eliminated the use of the gloves and is using Leap motion sensors to capture hand and finger movements. A recent study by Kardong-Edgren et al. (2019) shows that the VR game made nursing students feel engaged in their own learning and reported that it helped them learn how to insert a urinary catheter correctly (Kardong-Edgren et al., 2019).

According to Anthony Ellerson, Director of GIMM at Boise State University, his ultimate goal is to help nursing schools around the world use it and install them in university computer labs for any nursing students to use (He, 2018).

VR PROJECTS IN THE WORKS

Duke University School of Nursing

Duke University School of Nursing (DUSON), a private institution in Durham, North Carolina, is one of many schools that is currently developing its own application with the use of commercially available VR headsets. This new educational tool, being developed by DUSCON and the Durham VA Health Care System, is called "Task Trainer VR" (DUSON 360, 2018).

Dr. Michael Zychowicz, professor at DUSCON, began working on this project with Ethan Burch, a simulation fellow at the U.S. Department of Veterans Affairs, to help distance learners gain skills by practicing in VR. Task Trainer VR is built on existing technology, such as the Oculus Rift, so it can be ready to access from anywhere (DUSON 360, 2018).

A scenario within the Task Trainer VR is in a doctor's office with a simulated patient. Students will pick up a needle and be guided by lights on the patient, showing them where they need to inject the needle. If completed successfully, students will be rewarded with fireworks indicating that they correctly inserted the needle in the correct spot (DUSON 360, 2018). Though it is only in its early stages of development, it has great potential for distance learning in the near future for nursing students.

University of Michigan School of Nursing

In 2018, Oculus awarded nearly $100,000 in grant, along with four high-tech laptops and four sets of Oculus Rifts, to fund a VR pilot program led by Dr. Michelle Aebersold, Clinical Associate Professor at the University of

Michigan School of Nursing (UMSN), and Dr. Prashant Mahajan, department chair of pediatric medicine at Michigan Medicine. They partnered with the Duderstadt Center, located on the University of Michigan North Campus, to learn and build this program.

The goal of the program is to create an interactive, 3D environment where learners can engage in pediatric emergency care along with different types of healthcare professionals in the room. Following an algorithm for Pediatric Advanced Life Support (PALS), learners can practice the protocols, including using effective teamwork and communication skills to positively impact the child's outcome (Aebersold, n.d.).

The focus of this project is to develop, pilot, and evaluate a team-based PALS simulation by using the Oculus Rift to demonstrate the suitability of VR for improving communication, teamwork, PALS guideline, and patient outcome. The goal of this program is to reduce errors and improve the overall quality of care in the emergency department (ED) (Aebersold, n.d.).

Nursing students, who are often in an unpredictable environment such as the emergency department, will benefit from this program immensely and enhance their learning of pediatric emergency care and working with other healthcare professionals who responds to pediatric cardiac arrests (Aebersold, n.d.). Though the program is still being developed, it makes a strong statement that with the right resources, any school can develop its own in-house application.

IMPLEMENTING VR IN HIGHER EDUCATION

According to a recent 2017 survey, 65% of nursing programs are already using virtual simulation, while virtual reality will see an increase from 10% to nearly 45% in the next five years. This is a direct response to the worsening shortage of clinical training sites (Wolters Kluwer Health, 2017). Not only can VR assist with the lack of clinical training sites, but it could also be a huge cost saving to institutions. There are two big reasons why administrators and faculty in higher education should implement VR in nursing education: cost and customization.

Cost

The cost of medical simulation equipment is increasing each year, as old techs are phasing out and new equipment is being developed. When this occurs, institutions are forced to either use the equipment until it breaks, since support has ended for the product, or upgrade to the newest model. For example, a high-fidelity mannequin can cost up to $70,000 depending on what add-ons are chosen, such as a laptop or tablet, patient monitor, and other accessories. Warranty for these mannequins will generally only last a year.

Institutions who wish to maintain their simulators must purchase these warranties yearly, which can cost up to $7,000 per mannequin. For institutions that have simulation labs with over ten high-fidelity simulators, it could cost over $70,000 each year just for maintenance, not including sales tax. With VR, institutions can save thousands of dollars with little maintenance required compared to high-fidelity mannequins. For instance, the VR game developed by Boise State University only cost $8,000 to develop, including the equipment (He, 2018).

As VR technology becomes more prevalent, the price of headsets will start to decrease as well, making it more accessible to most everyone. Currently, some of the most expensive headsets cost approximately $5,000. Add a high-powered gaming PC or laptop, and institutions are looking at close to $8,000 per unit, a small price when compared to a high-fidelity mannequin. Some VR headsets do not even require a PC anymore and have plenty of power to run games wirelessly without the need for a dedicated graphics card.

Rather than purchase a mannequin for just one type of simulation, VR technology can allow institutions to create a multitude of scenarios to fit their specific needs. This will significantly reduce the cost of having to purchase multiple expensive mannequins. Institutions can save thousands by just investing in the software and letting the VR headset do the rest.

Customization

Unlike mannequins, the sky is the limit for VR technology. It allows for more customization in terms of scenario development and bringing the right moment to students at any time. Developers are already working hard each year to create new scenarios and games for healthcare professionals. For example, companies such as SimX quickly developed a VR COVID-19 training game for all nurses and healthcare workers to download for free, shortly after the pandemic began in the spring of 2020.

This was a huge advantage over mannequins, since any institution using VR can download this application instantly for free rather than re-creating an entire scenario from scratch. This provides students with the most up-to-date procedures that healthcare workers should be practicing.

VR technology has also created a professional network of developers who are committed to improving simulation-based learning and providing future healthcare providers with the latest resources needed to improve patient health outcomes. Demos can be downloaded electronically for institutions to test the product before committing. Once purchased, faculty can fully immerse their students in the game immediately once it is installed—another huge benefit over mannequins, which could take days just to schedule a site visit from a vendor. In addition, software can be shared with neighboring institutions or anyone as long as the application has been developed in-house.

Other companies such as UbiSim and Pocket Nurse have already started developing their own VR games as well. It is only a matter of time before large medical corporations such as Laerdal and CAE come out with their own products. Within a few years, many institutions will have the capacity to develop their own VR games, which will be unique to their nursing program. This will initiate an era of simulation-based learning with VR.

Administrators and faculty in higher education will need to quickly adopt this VR technology to prepare the next generation of nursing students. This will allow institutions to provide individuals with the best learning tools available to gain the necessary skills highly sought out by healthcare organizations and also save the institution thousands of dollars in equipment cost.

BENEFITS TO STUDENTS

Research shows that simulation-based learning is effective and engaging and provides an opportunity for students to apply the knowledge and skills they have learned in the classroom to practice. It is a great way to reduce anxiety and boost self-confidence for nursing students before they step out into the real world (Durham & Alden, 2008). With VR technology, students can feel confident and have the benefit of practicing clinical skills as much as they want in a safe environment. It is the best way to perfect their skills while reflecting and debriefing on what could be done to improve the scenario. This sets VR apart from simple task training.

VR technology is also entertaining, which engages individuals in experiential learning. Students learn better when they can see and do, which helps with retaining information. Students do not learn if they are not engaged. Even though VR is still in its infancy, the technology has allowed faculty to develop realistic scenarios to reflect a typical day of practice that nurses can expect. Many of these challenges can be overwhelming without the proper training. VR can help bridge this gap and add an extra benefit by developing scenarios to help nursing students improve in meeting patient needs and promoting patient safety.

Students growing up in the digital age will be more involved and eager to engage in these types of activities. The students of tomorrow may find it more exciting to learn in this way. VR technology will challenge the traditional method of teaching and replace certain aspects of the nursing curriculum. It may be the next educational method for students in the 21st century.

VR'S SUCCESS IN HIGHER EDUCATION

VR for nursing education is already here. It has changed the way some faculty are teaching and training future nurses in higher education. Not only has it already shown promise during its infancy stage, but it will evolve and get better as more research is done and advancements are made in technology.

Research already shows that students learn better through VR because it is fun and engaging. One of the reasons is because many students may already have VR headsets at home. Ten years ago, this was not the case. There were limited selections of games, the costs were extreme, and VR headsets were too bulky. Now, the technology has evolved with lighter VR headsets that produce better quality and are more comfortable to wear for prolonged periods. As the technology becomes more widely available, it is only natural that universities and community colleges should adopt it since students are already familiar with it.

VR has already established its place in nursing education as a viable option for many. Now, it is up to administrators and faculty to research and identify the benefits it could bring to the classroom/labs and the institutions.

POTENTIAL IMPACT ON HIGHER EDUCATION

Short Term

VR technology has already made its way into curriculum and clinical labs for some nursing programs. Students can complete homework assignments with VR and participate in virtual activities to enhance their understanding of the material. It has been the solution for replacing some clinical hours, such as during the COVID-19 pandemic in 2020. Institutions can expect students to be coming to universities and community colleges with a basic knowledge of the technology. It has already invaded their homes and personal devices, providing entertainment for millions of individuals.

VR has made substantial progress in simulation-based learning for students, while also meeting BON and INACSL standards for nursing simulation activities. In the near future, VR in nursing education will become more prevalent as high-fidelity mannequin and medical simulation equipment prices continue to rise. Institutions will look to VR as a viable alternative to providing a high level of immersive experiential learning for nursing students to improve patient health outcomes.

Mid Term

With large tech companies such as Apple, Google, Lenovo, Microsoft, and Facebook already heavily investing in VR, the technology will soon become available to virtually any individual who owns a smartphone in the next decade. Nursing programs may soon require a VR-capable device and a headset as part of their required supplies.

In a digital age, VR technology will become an essential tool for developing curriculum and online activities to further educate and enhance the learning experience for many online nursing programs. Faculty and staff will need to have a basic understanding of how VR works. Nursing programs may also require at least one faculty member to be an expert in VR for nursing education. As the demand for nursing staff increases, clinical training sites will continue to diminish, making it harder to secure clinical preceptors. Institutions must find a solution to prepare their students while also meeting business and industry workforce needs.

Long Term

Successful leaders in the future will need to start now and research the potential impact that VR will have on nursing education. In the next ten years, VR headsets will get dramatically smaller, and the need for controllers will cease. Instead, a full-body suit or gloves will become the new standard in providing haptic feedback to fully immerse individuals in virtual simulations. Applications will be able to support a massive number of players in one game, allowing for hundreds of individuals to participate in a simulated scenario and change the way nursing students communicate and interact with each other in the classroom.

Institutions must also be prepared for any mandates that the BONs may decide to implement in the next few decades. There may very well be requirements or options to replace up to 75% of clinical hours with simulation due to the continued shortage of clinical training sites that many nursing schools are facing. The future of VR looks bright, and many students continue to enjoy the learning approach by doing rather than seeing. As quantum computing comes into play, it will only be a matter of time before one will not be able to tell the difference between VR and reality.

CONCLUSION

As technology continues to change at an exponential rate, nursing continues to remain one of the leading professions to adopt these new technologies and incorporate them into their curricula. Institutions, such as universities and community colleges will need to research this technology to expand and continue providing support to their students and the local community. Though it does not entirely replace the emotional intelligence that many students get from traditional clinical sites, VR will still be a powerful tool that will change simulation-based learning and become the new pedagogy for the future in nursing education.

CHAPTER SUMMARY

- For years, nursing programs have been at the forefront of adopting and using healthcare simulation to provide a safe learning environment for nursing students before they enter the workforce.
- Over the years nursing programs have used a number of different mannequins such as Mrs. Chase, Resusci Anne, Sim One, and iStan to assist with training.
- Many of these high-fidelity mannequins are equipped with realistic physiologic responses including but not limited to, respiration, pulses, breath sounds, heart sounds, urinary output, pupil reaction, and ability to communicate to students.
- But why are so many nursing schools using simulation-based learning? One of the reasons is the decrease in the availability of clinical training sites.
- Simulation-based learning offers a unique approach to tackling the shortage of clinical training sites that many nursing schools are struggling to secure.
- Simulation-based learning has disrupted the way nursing programs have integrated simulation into their curriculum and clinical labs for decades.
- For a technology to qualify as VR, it must meet two main criteria in healthcare simulation. It must create a three-dimensional environment and must have an immersion effect.
- Many schools have jumped on the bandwagon of VR and the limitless possibilities that it can bring to the classroom and in clinical labs.
- Some examples of how VR is currently being used in nursing classrooms and clinical labs are Indiana University-Purdue University Columbus, San Jose State University, Penn State, Boise State University, Duke University, and the University of Michigan.
- The cost of medical simulation equipment is increasing each year as old techs are phasing out and new equipment is being developed.

- Rather than purchase a mannequin for just one type of simulation, VR technology can allow institutions to create a multitude of scenarios to fit their specific needs.
- VR technology has also created a professional network of developers who are committed to improving simulation-based learning and providing future healthcare providers with the latest resources needed to improve patient health outcomes.
- Research shows that simulation-based learning is effective and engaging and provides an opportunity for students to apply the knowledge and skills they have learned in the classroom into practice.
- Even though VR is still in its infancy stage, the technology has allowed faculty to develop realistic scenarios to reflect a typical day of practice that nurses can expect.
- Research already shows that students learn better through VR because it is fun and engaging.
- Though it does not entirely replace the emotional intelligence that many students get from traditional clinical sites, VR will still be a powerful tool that may disrupt simulation-based learning and become the new pedagogy for the future in nursing education.

REFERENCES

AACN fact sheet: Nursing shortage. (2017, May 18). Retrieved May 2, 2020, from https://www.aacnnursing.org/Portals/42/News/Factsheets/Nursing-Shortage-Factsheet-2017.pdf?ver=2017-10-18-144118-163&ver=2017-10-18-144118-163

Aebersold, M. (n.d.). Research. Retrieved May 23, 2020 from https://sites.google.com/a/umich.edu/michelle-aebersold/research-projects

Appelman, H. (2018, July 24). World Campus researches effectiveness of VR headsets and video in online classes. Retrieved May 15, 2020, from https://news.psu.edu/story/529049/2018/07/25/research/world-campus-researches-effectiveness-vr-headsets-and-video-online

Bradley, C. S., Johnson, B. K., Dreifuerst, K. T., White, P., Conde, S. K., Meakim, C. H., Childress, R. M. (2019). Regulation of simulation use in United States prelicensure nursing programs. *Clinical Simulation in Nursing, 33*, 17–25. doi:10.1016/j.ecns.2019.04.004

Bratton, J. (2019, June 3). Virtual reality nursing. Retrieved May 9, 2020, from https://sciencenode.org/feature/Preparing%20for%20real%20life%20with%20VR.php

Chau, B. (2017, June 06). IMedicalApps: Will VR replace traditional medical training? Retrieved on May 15, 2020, from https://www.medpagetoday.com/Blogs/IltifatHusain/65815?xid=nl_mpt_DHE_2017-06-07

Cooper, J. B., & Taqueti, V. R. (2008). A brief history of the development of mannequin simulators for clinical education and training. *Postgraduate Medical Journal, 84*(997), 563–570. doi:10.1136/qshc.2004.009886

Durham, C., & Alden, K. R. (2008, April). Enhancing patient safety in nursing education through patient simulation. In: Hughes, R. G., editor. Patient safety and quality: An evidence-based handbook for nurses. Rockville (MD): Agency for Healthcare Research and Quality (US). Chapter 51 Retrieved on April 25, 2020, from https://www.ncbi.nlm.nih.gov/books/NBK2628/

DUSON 360. (2018, November 20). Virtual Reality Task Trainer - DUSON 360 Episode #22 [video]. YouTube. https://www.youtube.com/watch?v=w7_q3FaDBFo&feature=emb_title

He, B. (2018, April 7). Nursing students are learning medical techniques with VR. Retrieved on May 24, 2020, from https://arvrjourney.com/nursing-students-are-learning-medical-techniques-with-vr-fd4c5a8d642d

Kardong-Edgren, Suzan, PhD, RN, ANEF, CHSE, FSSH, FAAN, Breitkreuz, Karen, EdD, MSN, et al. (2019). Evaluating the usability of a second-generation virtual reality game for refreshing sterile urinary catheterization skills. *Nurse Educator, 44*, 137–141. https://doi.org/10.1097/NNE.0000000000000570

Lioce L., Lopreiato J., Downing D., Chang T. P., Robertson J. M., Anderson M., Diaz D. A., and Spain A. E. (Assoc. Eds.) and the Terminology and Concepts Working Group (2020), Healthcare simulation dictionary–second edition. Rockville, MD: Agency for Healthcare Research and Quality; January 2020. AHRQ Publication No. 20-0019. DOI: https://doi. org/10.23970/simulationv2.

Madrid, C. (2016, April 24). Boise State video "game" could transform nursing education. Retrieved May 24, 2020, from https://www.idahostatesman.com/living/health-fitness/article73710932.html

Merriam-Webster. (n.d.). Three-dimensional. In *Merriam-Webster.com dictionary*. Retrieved May 15, 2020, from https://www.merriam-webster.com/dictionary/three-dimensional

Modeling and Simulation (M&S) Glossary. (2016). Retrieved May 15, 2020, from https://www.msco.mil/MSReferences/Glossary/TermsDefinitionsI-M.aspx

SSH. (2020). Society for simulation in healthcare. Retrieved April 25, 2020, from https://www.ssih.org/

Syminar. (2018). Retrieved June 19, 2020, from https://www.syminar.com/

Weir, W. (2012, March, 29). At 101, "Mrs. Chase" is a medical marvel. Retrieved from April 25, 2020, from https://www.courant.com/health/hc-xpm-2012-03-29-hc-mrs-chase-hartford-hospital-0328-20120327-story.html

Wolters Kluwer Health. (2014, January 27). *Laerdal medical and Wolters Kluwer Health introduce virtual simulation learning tool for nursing students* [press release]. Retrieved June 5, 2020, from http://download.lww.com/vsim/VSim_PressRelease_012714final.pdf

Wolters Kluwer Health. (2017, September 14). National nursing survey shows adoption of new technologies in educational programs to enhance nurse graduate practice readiness. Retrieved June 5, 2020, from http://home.lww.com/news.entry.html/2017/09/14/national_nursingsur-rZTw.html

Chapter 2

Anatomy and Physiology

Candice B. Lewis

Human anatomy and physiology (A&P) are related disciplines that span the educational spectrum. According to the Merriam-Webster online dictionary, anatomy "is the art of separating the parts of an organism in order to ascertain their position, relations, structure, and function" ("Anatomy," n.d.).

Physiology is defined by the Merriam-Webster online dictionary as "a branch of biology that deals with the functions and activities of life or of living matter (such as organs, tissues, or cells) and of the physical and chemical phenomena involved" ("Physiology," n.d.). Children learn basic human A&P early and continue to learn throughout school and into college. Those entering medical professions will continue more in-depth studies of A&P as part of the field of study.

Before moving into the future of A&P education, it is important to understand a brief history of how the study and teaching in each field has developed. According to Bisht, Hope, and Paul (2019) the earliest records indicating drawings and notes on human anatomy were found dating back to 1600 BCE. Historical records begin to include reference to medicine just 50 years later (Bisht, Hope, & Paul, 2019).

It is believed these early records of A&P began with physical examination of bodies postmortem in early Greece. In 1200 BCE, records from India also indicate surgical notes and records of medicines (Bisht et al., 2019). In addition, Chinese medical records from around 400 BCE indicate knowledge of normal aging effects on the body's A&P (Bisht et al., 2019). Other nations continued to add to the recorded knowledge of A&P later in history.

EARLY ANATOMY AND PHYSIOLOGY EDUCATION

Early teaching in medical education was accomplished through apprenticeships (Gourevitch, 1999). Eventually, these apprenticeships developed into formal medical schools and texts containing A&P began to be printed in the mid-1400s. These concurrent events helped to establish the formalized study of A&P (Bisht et al., 2019).

Many scholars and early scientists contributed to the foundation of the fields of A&P. The study of these subjects is believed to have begun by examination of postmortem specimens and by referencing the drawings and notes of others (Bisht et al., 2019). According to Standring (2016) the individual considered the "father of modern anatomy" is Andreas Vesalius (p. 8).

Vesalius lived during the first half of the 16th century and created detailed illustrations and notes on anatomy. He was not the first individual to create drawings; rather, he noted where previous drawings were not accurate and provided more reliable drawings and notes (Standring, 2016). Within the next century, tools and methods of study began to allow for microscopic examination, leading to a greater understanding of disease of the body (Bisht et al., 2019).

Anatomical Models

The history of A&P education is one that also involves the influence of artists. A modern student of A&P has likely used a model displaying details of a human cell or the entire human digestive system created by an artistic replication of a drawing. These models began development in the 17th century and have been used for generations of student instruction (Marković & Marković Živković, 2010).

Just as interest in A&P grew, so did interest in methods used to display the human body. The earliest tools were drawings and notes, these were followed by visual models. Anatomical models were first used in China, and the practice of creating models soon reached Italy, Germany, and other countries throughout the world. Early anatomical models were formed from wood, silk, ivory, and papier-mâché. Specimen preservation also helped in the instruction of anatomy. The models and preserved specimens allowed for a reduced reliance on human cadavers (Marković & Marković, Živković, 2010).

TWENTIETH CENTURY ANATOMY AND PHYSIOLOGY INSTRUCTION

Throughout the 20th century, students benefited from medical drawings illustrating human anatomy and physiologic processes. Students in K–12 settings also gained experience in learning from models and sometimes are asked to create models as part of instruction. In some settings, students may also have watched films demonstrating A&P principles.

Students in the college setting, especially the late 20th century, most likely used images of cadaver dissections and sectional images from computed tomography (CT) and magnetic resonance imaging (MRI) studies from the 1990s (Standring, 2016). As students advanced to postgraduate education in the study of medicine, schools provided cadaver dissection as a traditional part of the training for medical degrees (Drake, McBride, Lachman & Pawlina, 2009).

TWENTY-FIRST CENTURY ANATOMY AND PHYSIOLOGY INSTRUCTION

The study of A&P today is supported through the advancement of computing. Additionally, it is important to note that medical imaging developments and advancements in CT and MRI have also contributed to 21st-century tools. These modalities contribute by providing images from patient studies that can be converted to a 3D image and then be manipulated by a user to perceive the height, width, and length of the image subject (Chien, Chen & Jeng, 2010).

There are four specific areas to consider regarding emerging instructional tools that are made possible by computing technology related to instruction in A&P: three-dimensional (3D), virtual reality (VR), augmented reality (AR), and mixed reality (MR).

Three-Dimensional

3D anatomy and physiology programs allow for users to view a display of anatomy or a physiological process. 3D models can be handheld models (such as a model of a heart) that can be physically manipulated by a student. A 3D model can also be generated digitally and used in a VR or AR system to be displayed and manipulated (Fredieu, Kerbo, Herron, Klatte & Cooke, 2015).

3D provides an advantage over a two-dimensional (2D) image or 2D computer display. A 2D image will only allow for depictions of the height and width of a structure, thus the user cannot perceive the depth of the structure

(Fredieu et al., 2015). In short, 2D images may limit the learner to understanding the spatial relationship of structures in the body.

A&P instruction in higher education has long depended on 2D images in textbooks and handheld models. For example, the Visible Human Project in the mid 1990s provided images based on actual CT and MRI cross sections of both an entire male and female cadaver. These images could then become part of gross anatomy instruction and created opportunities for the images to be reflected as 3D images (Bisht et al., 2019).

Anatomage (2018) is a product that projects a human cadaver onto a horizontal, table-size platform. This platform allows for virtual dissection of the cadaver while being able to rotate the cadaver to view from various planes. There are additional products available through Anatomage for the study of A&P. According to the website for Anatomage, the company is currently in developing and launching of an e-book format for student use at home.

3D computer models can also be found as supplemental features for textbooks from publishing companies such as McGraw-Hill (2020). Supplemental 3D images are also available for students to download independently through applications for purchase. These applications are gaining wider acceptance as various levels of higher education are adapting this technology to A&P instruction (K. Haynie, personal communication, June 24, 2020).

Virtual Reality

VR is defined as enabling the user to feel part of the display. VR can be displayed through an Oculus device or in a structure of surround screens in which the user wears a tracking device. The device assists the image to move and respond to the user's movements (Fairén, Farrés, Moyés & Insa, 2017). There are several examples of this technology now on the market (R. Duran, personal communication, July 15, 2020).

An example of VR technology is the system created by Sharecare (2020), which features simulated A&P. This system runs on a variety of viewers or head-mounted devices (HMD). A second example is The BodyVR, which uses actual CT, MR, and other medical imaging to create a virtual experience for students to be immersed in A&P (The Body VR, 2020). This product is used with a head-mounted display provided by one of the following companies: Google AR & VR (2020), HTC (2020), and Oculus (n.d.). A third example is the 3D Organon Virtual Anatomy (2020), which uses Oculus Rift (n.d.) and HTC (2020).

Mixed Reality and Extended Reality

Mixed reality (MR) is a term used to refer to the combination of virtual and non-virtual environments to deliver content (Virtual Reality & 360 Video Services, 2020). An example of this technology in use is found with the Microsoft HoloLens (2020). This product, like other HMDs, is not limited to educational use, but works well for applications designed for A&P study. The product is being used in A&P education at Case Western Reserve University, Cleveland Clinic medical education programs (University News, 2019, April 9).

According to the University of New Mexico website's glossary of terms (Virtual Reality & 360 Video Services, 2020):

> Extended Reality (XR) can be defined as an umbrella, which brings all three Realities (VR, AR, and MR) together under one term. XR refers to all real-and-virtual combined environments and human-machine interactions generated by computer technology and wearables. (para. 4).

According to Zweifach and Triola (2019), XR is poised to continue growing in A&P education as technology improves and the opportunity to replicate the human body and physiologic functions.

WHERE AND HOW: VR FOR ANATOMY AND PHYSIOLOGY IN PRACTICE

Modalities for teaching A&P are improving rapidly. This segment will include community colleges, four-year colleges/universities, and medical schools. Just as the modalities are rapidly being refined, vendors are working with higher education to meet the needs of the changing education environment. According to Pantelidis et al. (2018), A&P lends itself easily to VR modalities when compared to other subjects. Concannon, Esmail, and Roduta Roberts (2019) report that in health science education, surgical and anatomy training are the second and third most common VR programs, respectively. It will be interesting to see how these technologies continue to grow in the future.

Community Colleges

A&P courses at the community college level are often prerequisite courses required for students entering the health field whether in allied health or nursing programs. For this review of the use of VR in the community college setting, instructors and VR lab directors at five community colleges were contacted. Each provided information on any virtual tools being used for course delivery. At the time of this writing, the COVID-19 pandemic was in its fourth month in the United States. This is important as a consideration for the use of VR in A&P courses as it created challenges for equipment held on college campuses during a move to online learning environments.

Chattahoochee Valley Community College

Chattahoochee Valley Community College (CVCC) in Phenix City, Alabama, opened two computer labs featuring zSpace computer workstations, which feature VR technology for the study of A&P. The labs were made possible through a grant for a total of 24 stations, each station featuring a HMD. One important consideration for obtaining the labs at CVCC was that the local high schools were using similar technology, thus creating a pipeline of students familiar with the technology.

Dr. Jackson at CVCC stated that the VR lab used for instruction in A&P provided an opportunity for students to gain a "better understanding of the organ system and the area of the body being referenced." Dr. Jackson expressed that the VR tool allows for students to conduct dissections as well as opportunities for the visualization of a disease process (B. Jackson, personal communication, June 9, 2020).

Mississippi Gulf Coast Community College (MGCCC)

Another example of a community college using VR is MGCCC. The college has developed an on campus lab which features the Oculus Rift (n.d.) and zSpace (2020) computer workstations that can be used for instruction in A&P. In addition, the library at MGCCC loans Oculus Rift (n.d.) to students. (K. Rouse, personal communication, June 9, 2020).

In addition to the VR available at MGCCC, the college also added AR for use in A&P. Dr. Kelly Rouse, MGCCC science instructor, described in a personal interview that the use of AR through Visible Body (2020) has allowed students in A&P courses to have access whether on campus or from home. This product does require students to download the application and any updates while connected to the college network.

In particular, Dr. Rouse found the ability of students to have this application for home use was very useful during the transition to online instruction during the pandemic in spring 2020. Dr. Rouse uses the technology during class and has found that students express the use of AR is engaging. Dr. Rouse has noted the system has appeared to make learning the skeletal structures easier for students. A key feature is being able to project a virtual cadaver to guide students through the body from multiple planes. Students are able to replicate this experience through use of personal devices from home or another remote location (K. Rouse, personal communication, June 9, 2020).

Houston Community College

Houston Community College (HCC) opened its first virtual reality lab in 2016 on the Central Campus. The college also added another VR lab at the West Loop campus in January 2020; however, the lab on the Central Campus is where applications for A&P are offered (R. Duran, personal communication, July 15, 2020). The labs are used to support several programs and courses throughout the college, including A&P.

HCC's Center for Learning Technology has guided the development of the VR labs. With the first lab in 2016, HCC was one of the first community colleges to feature VR. Since this time, HCC has been committed to continuing to grow VR and AR. The college enjoys partnerships with local development companies for continuing expansion.

HCC currently has 10 units in a lab setting at the Central Campus. The units each have a HMD and a 55-inch display screen for other students to view in small groups. The primary application used is Sharecare (2020). Mr. Duran has found faculty have learned the technology best in groups and become more comfortable.

Mr. Duran indicated that faculty have moved to adopting the technology for instruction in A&P. An example provided for how the technology is incorporated is through the instructor manipulating the anatomy and then making annotations for students. These sessions are recorded and then uploaded to the HCC learning management system (LMS) for students to review as an assignment.

In regards to student engagement, Mr. Duran reports student surveys reveal a high level of satisfaction, and students report they retain material better using VR compared to traditional lectures and labs. Students in online classes also use the lab by coming in and completing assignments. These assignments can then be documented and submitted through the LMS for instructor review (R. Duran, personal communication, July 15, 2020).

Four-Year Colleges and Universities

VR appears to be gaining popularity at the four-year college and university level at a faster pace compared to the community college. Several schools were found to have installed and implemented VR labs. Examples are East Carolina University (Dilda, 2019), Jackson State University (Warren, 2019), Nova Southern University (2020), Western Carolina University

(WCU, Cantrell, 2020), Mississippi State University (Carskadon, 2017), and many others.

In particular, Pennsylvania State University is using the technology for all students at the university. For accessibility, students can view available VR applications through the university web page using recommended accessories for viewing (Penn State, 2020). Of note is that some four-year institutions are not only using the various modalities of virtual A&P, but they are also creating software and hardware customized to meet the needs of the institution for teaching A&P for both undergraduate and graduate education.

Examples of use can be found at Colorado State University (2020). CSU developed an anatomy VR program through cooperation of the anatomy department and the university's virtual reality Initiative. Cornell College in Iowa developed a VR surgical planning program in 2017 through cooperation of the engineering program.

Although many colleges and universities are adapting and employing VR for A&P instruction, some institutions are finding the 3D technology also meets instructional needs. Dr. Victoria Turgeon at Furman University instructs both undergraduate and graduate students using a 3D virtual anatomy system, Bodyviz (n.d.). The tool is used in various anatomy courses to support both normal and abnormal anatomy studies. One attraction of this type of system is the use of CT and MR studies for review by students. Dr. Turgeon uses this feature because access to CT and MR images of anatomy and pathologies are not readily possible through other sources (V. Turgeon, personal communication, June 13, 2020).

Medical Schools

Medical schools have traditionally relied on human cadaver dissection for instruction in gross anatomy (Drake et al., 2009). Cadavers present challenges for medical schools such as storage, cost, and availability (Singh & Kharb, 2013). VR, AR, and MR have been used within medical schools, and this use is not without debate on efficacy. Medical schools have debated on a reliance on cadavers and some medical schools are beginning to use VR (McMenamin et al., 2018).

Medical schools have been adopters of VR systems for teaching A&P and associated skills. The reduced time spent on instruction in anatomy may be one reason medical schools have been willing to adopt this technology (Chien et al., 2010). Based on a review of the literature on the use of VR, such technology is used most often as a complement to instruction with cadavers. The focus of instruction is more on anatomy than physiology.

An advantage of using VR to support anatomy education is that the student can move from the cadaver lab to a lab supporting VR to continue study. In the case of schools such as Penn State (2020), students can also use these tools at home. VR has also been applied in surgical planning (Cornell College, 2017).

Conducting surgery for repair or joint replacements using a VR approach allows surgeons and medical students to practice using surgical equipment while practicing on a virtual patient (Meyer, James, & Cui, 2018). A second area is in learning structures that are very small and difficult to locate on a cadaver. An example is in the inner ear, which is composed of small bones (Nicholson, Chalk, Funnell, & Daniel, 2006).

The majority of medical schools are continuing to use cadavers as a basis of anatomy and are adding VR to enhance and supplement training (Heather, Chinnah, & Devaraj, 2019; Zweifach & Triola, 2019). Although research is underway to determine the effectiveness of this model, there are some schools moving away from cadavers.

An example of a medical school moving away from cadavers is found at Case Western Reserve University. Case Western and the Cleveland Clinic completed a new building in 2019 which joins medical, dental, and nursing education in one building. As part of the new facility, the medical program also implemented VR anatomy labs in place of cadaver labs. These labs can be shared among programs that are colocated in the new facility (University News, 2019).

VR IN HIGHER EDUCATION: BENEFITS

Programs in VR for use by various levels of education have become more and more accessible. Academic publishers of textbooks are also providing students with virtual learning tools that can be used at home. Companies have partnered with publishers to provide in-class equipment that creates a 3D VR for use on the student's home computer or mobile device. Other applications provide students the opportunity to download tools that can be used at home and in turn create opportunities for students to study detailed images at school, at home, and even while traveling (K. Rouse, personal communication, June 9, 2020).

The flexibility created by the portability of VR was evident during the period of the pandemic associated with COVID-19. In spring 2020, many colleges moved all instruction to an online format in an effort to reduce the rate of spread of COVID-19 (Burke, 2020). As previously stated, community colleges were able to move instruction online with the available VR tools.

Western Carolina University (Cantrell, 2020) and Case Western Reserve University/Cleveland Clinic (University News, 2019, April 9) are examples of two university-level institutions that were able to move online using VR tools to continue instruction in A&P courses. These options create more opportunities for student users to access material for study and practice.

Spatial Resolution

Falah et al. (2014) address the challenges of learning spatial relationships through a traditional 2D format: "In addition, a spatial relationship understanding from traditional modalities (textbooks and 2-dimensional [2D] images) is not clear; requires expertise to explicate, and lacks adequate detail to demonstrate" (p. 371). According to Chien et al. (2010), VR appears to assist with improvement of spatial understanding for learners by demonstrating the relationship of anatomical structures throughout the body.

Zweifach and Triola (2019) state that the use of head-mounted devices (HMD) in the VR setting allow students to become spatially aware of organs and structures. Nicholson et al. (2006) demonstrated this feature when using the tool for study of the inner ear and the unique spatial relationship of this particular anatomical relationship.

Student Engagement

Although there are still many questions surrounding the effectiveness of the use of VR for teaching, there are indications of a benefit for learners. According to Falah et al. (2014) "real-time VR applications offer a rich, interactive, and highly engaging educational context, thus supporting experimental learning-by-doing" (p. 372). The ability of colleges and universities to offer the applications through libraries, VR labs, and student devices provides additional opportunities for students to review material outside a set lab or lecture time.

Several studies reviewed also indicated a high rate of satisfaction for students using VR. Students reported finding the experience of learning A&P as engaging (Heather et al., 2019). Chen et al. (2019) found using the Microsoft HoloLens compared to physical models showed the models to be more effective based on student performance in memory recall compared to those using HoloLens. However, the students were more engaged and had greater satisfaction in the learning experience using the HoloLens.

Benefits for Colleges

With moving toward or fully adopting VR, colleges may be able to reduce costs. For example, the use of VR products by colleges and medical schools may reduce costs associated with purchases of models and cadavers (Zweifach & Triola, 2019). Fredieu et al. (2015) also agree that A&P laboratories based on models present challenges for colleges in terms of cost of replacing heavily used models that may become damaged or misplaced.

Just as publishers are adding 3D tools that can be used on a student's home computer or mobile device, this same trend is also present already with VR programs (B. Jackson, personal communication, June 9, 2020). This may not completely eliminate the physical lab, but in the events surrounding COVID19, it does demonstrate that this format may be able to provide a continuum of learning for students beyond physical labs.

CONSIDERATIONS AND LIMITATIONS

There are several benefits to the use of VR for studying A&P for college students. Although the technology is becoming more widely adopted throughout higher education, limitations have been identified. The common limitations will be explored further in this section of the chapter. "However, uptake of the technology is still limited. This can be due to concerns regarding the purchase cost and software development, as well as uncertainty over usability or efficacy of the technology" (Heather et al., 2019, p. 12).

In a personal interview with Ashley Bondurant at Trident Technical College in Charleston, South Carolina, several concerns regarding the technology were shared. One concern shared was the initial cost of of adopting the technology. Another concern was the continued need for research to demonstrate the educational efficacy of VR. There are also concerns regarding Oculus sickness or cybersickness (A. Bondurant, personal communication, June 9, 2020).

Cybersickness

Cybersickness is characterized by Gallagher and Ferrè (2018) as being similar to motion sickness and may cause a user to experience nausea, disorientation, and tiredness. It is reported to be associated with the use of a HMD during a VR experience. This sensation may not affect all users the same, and being prone to this effect may be associated with gender, age, and the length of time a HMD is used in the VR environment (Gallagher & Ferrè, 2018). "According to the sensory-mismatch theory, the most likely cause of cybersickness is a mismatch between visual stimuli and the appropriate vestibular or proprioceptive feedback" (Gavgani, Walker, Hodgson, & Nalivaiko, 2018, p. 1671).

Although there is no definitive way to predict who may be impacted by cybersickness, individuals who are more prone to motion sickness are found to be more prone to cybersickness as well. Motion sickness and cybersickness sensations are similar and share many common, uncomfortable effects on an individual (Gavgani et al., 2018).

Repeated exposure to VR may reduce cybersickness, but this may take several exposures. In addition, when used for prolonged periods, HMDs may cause some users to continue to feel movement sensations even after discontinuing use (Gallagher & Ferrè, 2018). These are factors that instructors and colleges should recognize so users can be made aware for safety reasons.

Efficacy of VR

VR has been shown to be a complement to the reduced hours offered in the study of gross anatomy in the medical school setting (Chien et al., 2010). Although many universities and community colleges are adopting the technology to teach A&P, the question of the technology being at a point where it is a greater educational teaching tool compared to standard practices remains a debate.

An example was in findings from Lombardi et al. (2014) related to student performance when comparing the use of models, physical dissection, and virtual dissection related to heart A&P. Students using the models were found to perform better on a follow-up exam when compared to students using physical dissection or virtual dissection tools. However, one limitation of this study was the small sample size of 29 students (Lombardi et al., 2014).

One caution concerning the use of VR for physiology is the potential for programs to oversimplify concepts for learners. Challenging concepts related to A&P may require instructors to support student learning beyond VR (Richardson, 2011). As Heather et al. (2019) conclude, more studies are needed to discern the benefits of VR to student learning related to A&P.

Cost Concerns

Cost is a concern when investing in VR. In speaking with community colleges, this concern was overcome through grant writing in some cases (B. Jackson, personal communication, June 9, 2020). Colleges wishing to adopt this technology must take cost into consideration. In the case of AR software that is downloaded on a student's personal device, the software may require the student to invest in a device with sufficient memory and the capacity to handle the software and subsequent updates. This requirement can be an added cost to the student. As Heather et al. (2019) indicate, "Further research is required into the depth and longevity of learning and the cost-effectiveness of VR" (p. 15).

Learning Curve

As mentioned previously, some technology does have a time-demanding learning curve for both students and instructors. For some educators this learning curve may be daunting and impact adoption (K, Rouse, personal communication, June 9, 2020). When planning for courses, faculty will have to plan ahead to become familiar with the technology. In addition, students may struggle and need assistance. It was noted on several websites that colleges and universities have provided instructor guidance as part of using VR technology for instruction: (R. Duran, personal communication, July 15, 2020).

WHY VR WILL WORK AT THE COMMUNITY COLLEGE

There is certainly debate on the benefits of VR. One consistent finding is that students indicate a high satisfaction when using VR in learning A&P (Heather et al., 2019). VR appears to encourage engagement with little effort. Heather et al. (2019) conclude "It can be observed that VR . . . had a notable impact on spatial understanding, motivation and interest. Other reported areas of benefit included autonomous learning, confidence, cognitive load and overall student perceptions" (p. 4).

Another consideration for why VR can work at the community college level is the availability of the technology is becoming more widespread—this is evident in the number of products available for the various types of VR. In addition, the supporting devices like HMD are becoming more cost effective with the availability of products such as Google Cardboard, which can be used with a student's mobile phone.

The variety of products in VR have become more and more realistic (Meyer et al., 2018). This realism may never replace cadaver dissection, but it provides a complement to traditional methods of teaching A&P. Fairén et al. (2017) propose "Virtual Reality tends a bridge between what can be imagined in 2D and the 3D reality" (p. 58).

Finally, with the impact of COVID-19 and the need to move courses mostly to an online environment, VR provided an excellent option for conducting A&P labs online for colleges with the technology on hand (Cantrell, 2020). The movement toward online delivery may spur more colleges to consider the adoption of VR for A&P.

Imagine a community college or university student taking A&P sitting at home and being able to project a human body into a room and then be able to study the full digestive tract by removing the organs and then placing them back in order in the body cavities. The program then conveys an accuracy level and suggests hints to remember the order of the small intestine segments. This scenario is very different from the classic use of colored 2D images and sitting in a laboratory taking apart a model while a single instructor supervises a room full of students.

Although VR is not limited to just A&P, it has provided many resources for students to study and learn about the topic. Medical schools, colleges, and high schools are using the technology and showing that the tools are engaging for student learners (Bisht et al., 2019). As the technology continues to increase in realism, availability, and affordability, the use of VR will become more normalized across the education continuum.

CONCLUSION

The future of VR in A&P is continuing to evolve in exciting ways. Both students and educators will benefit from this technology and be part of its exciting future. Considering that students have been shown to enjoy VR, that VR is effective for learning, and that the technology is portable, colleges not considering the addition of VR may find themselves behind and working to catch up with the technology curve.

CHAPTER SUMMARY

- Before moving into the future of A&P education, it is important to understand a brief history of how the study and teaching in each field has developed.
- Many scholars and early scientists contributed to the foundation of the fields of A&P.
- The history of A&P education is one that also involves influence of artists.
- Specimen preservation also helped in the instruction of anatomy.
- The study of A&P today is supported through the advancement of computing.
- There are four specific areas to consider regarding emerging instructional tools that are made possible by computing technology related to instruction in A&P: three-dimensional (3D), virtual reality (VR), augmented reality (AR), and mixed reality (MR).
- 3D anatomy and physiology programs allow for users to view a display of anatomy or a physiological process.

- VR is defined as enabling the user to feel that they are part of the display. VR can be displayed through an Oculus device or in a structure of surround screens in which the user wears a tracking device.
- For this review of use of VR in the community college setting, instructors and VR lab directors at community colleges were contacted.
- They are Chattahoochee Valley Community College, Mississippi Gulf Coast Community College, and Houston Community College.
- VR appears to be gaining popularity at the four-year college and university level at a faster pace compared to the community college.
- Examples of use can be found at the Colorado State University, Cornell College, and Furman University.
- Medical schools have been adopters of VR systems for teaching A&P and associated skills.
- An example of a medical school moving away from cadavers is found at Case Western Reserve University.
- The flexibility created by the portability of VR was evident during the period of the pandemic associated with COVID-19.
- The ability of colleges and universities to offer the applications through libraries, VR labs, and through student devices provides additional opportunities for students to review material outside of a set lab or lecture time.
- With moving toward or fully adopting VR, colleges may be able to reduce costs.
- Although the technology is becoming more widely adopted throughout higher education, limitations have been identified.
- Cybersickness is characterized by Gallagher and Ferrè (2018) as being similar to motion sickness and may cause a user to experience nausea, disorientation, and tiredness.
- One caution concerning the use of VR for physiology is the potential for programs to oversimplify concepts for learners.
- Cost is a concern when investing in VR.
- Although VR is not limited to just A&P, it has provided many resources for students to study and learn about the topic.

REFERENCES

3D Organon Virtual Reality (2020). Retrieved from https://www.3dorganon.com/applications/3d-organon-vr-anatomy/

Anatomage (2018). Retrieved from https://www.anatomage.com/anatomage-medical/

Anatomy (n.d.). *In Merriam-Webster.com dictionary*. Retrieved from https://www.merriam-webster.com/dictionary/anatomy

Azer, S. A., & Azer, S. (2016). 3D anatomy models and impact on learning: a review of the quality of the literature. *Health professions education*, *2*(2), 80–98. doi.org/10.1016/j.hpe.2016.05.002

Bisht, B., Hope, A. & Paul, M. K. (2019). From papyrus leaves to bioprinting and virtual reality: History and innovation in anatomy. *Anatomy & Cell Biology 52*(3), 226–235. doi:10.5115/acb.18.213

BodyViz (n.d.). Retrieved from https://www.bodyviz.com/

Burke, L. (2020, March 9). Colleges move classes online as coronavirus infects more. Retrieved from https://www.insidehighered.com/print/news/2020/03/09/colleges-move-classes-online-coronavirus-infects-more

Cantrell, G. (2020, April 27). Hunter Library continues to assist faculty, students with virtual research, including 3D anatomy. Retrieved from https://news-prod.wcu.edu/2020/04/hunter-library-continues-to-assist-faculty-students-with-virtual-research-including-3d-anatomy/

Carskadon, J. (2017, November 10). MSU libraries, CAVS collaborate to provide students with access to virtual and augmented reality. Retrieved from https://www.msstate.edu/newsroom/article/2017/11/msu-libraries-cavs-collaborate-provide-students-access-virtual-and

Chen, C., Zhang, L., Luczak, T., Smith, E., and Burch, R. F. (2019). Using Microsoft HoloLens to improve memory recall in anatomy and physiology: A pilot study to examine the efficacy of using augmented reality in education. *Journal of Educational Technology Development and Exchange, 12*(1), 17–31. doi: 10.18785/jetde.1201.02

Chien, C. H., Chen, C. H., & Jeng, T. S. (2010). An interactive augmented reality system for learning anatomy structure. *Proceedings of the International Multi-Conference of Engineers and Computer Scientists*, *Hong Kong*: 1, 17–19.

Colorado State University (2017). Virtual reality lab at Colorado State University. Retrieved from https://www.youtube.com/watch?v=zDrLMgYZcac.

Concannon, B. J., Esmail, S., & Roduta Roberts, M. (2019). Head-mounted display virtual reality in post-secondary education and skill training: A systematic review. In *Frontiers in Education 4,* 80. doi:10.3389/feduc.2019.00080

Cornell College (2017). Research team creates virtual reality surgical simulator. Retrieved from https://news.cornellcollege.edu/2017/10/research-team-creates-virtual-reality-surgical-simulator/

Dilda, K. R. (2019, October 2). Virtual reality comes to ECU: Laupus library unveils first virtual reality lab for campus community. Retrieved from https://news.ecu.edu/2019/10/02/virtual-reality-comes-to-ecu/

Drake, R. L., McBride, J. M., Lachman, N., & Pawlina, W. (2009). Medical education in the anatomical sciences: The winds of change continue to blow. *Anatomical Sciences Education*, *2*(6), 253–259.

Fairén, M., Farrés, M., Moyes, J., & Insa, E. (2017). Virtual reality to teach anatomy. *Eurographics*. 51-58. doi:10.2312/eged.20171026

Falah, J., Charissis, V., Khan, S., Chan, W., Alfalah, S. F., & Harrison, D. K. (2014, August). Development and evaluation of virtual reality medical training system for anatomy education. In *Science and Information Conference* (pp. 369–383). Springer, Cham. DOI 10.1007/978-3-319-14654-6_23

Fredieu, J. R., Kerbo, J., Herron, M., Klatte, R., & Cooke, M. (2015). Anatomical models: A digital revolution. *Medical Science Educator*, *25*(2), 183–194. doi:10.1007/s40670-015-0115-9

Gallagher, M., & Ferrè, E. R. (2018). Cybersickness: A multisensory integration perspective. *Multisensory Research*, *31*(7), 645–674. doi:10.1163/22134808-20181293

Gavgani, A. M., Walker, F. R., Hodgson, D. M., & Nalivaiko, E. (2018). A comparative study of cybersickness during exposure to virtual reality and "classic" motion sickness: are they different? *Journal of Applied Physiology*, *125*(6), 1670–1680. doi:10.1152/japplphysiol.00338.2018.

Google AR & VR (2020). Retrieved from https://arvr.google.com/

Gourevitch, D. (1999). The history of medical teaching. *The Lancet*, 354, SIV33.

HTC (2020). Retrieved from https://www.htc.com/us/

Heather, A., Chinnah, T., & Devaraj, V. (2019). The use of virtual and augmented reality in anatomy teaching. *MedEdPublish*, *8*.

Lombardi, S. A., Hicks, R. E., Thompson, K. V., & Marbach-Ad, G. (2014). Are all hands-on activities equally effective? Effect of using plastic models, organ dissections, and virtual dissections on student learning and perceptions. *Advances in physiology education*, *38*(1), 80–86. doi:10.1152/advan.00154.2012

McGraw-Hill (2020). Retrieved from https://www.mheducation.com/highered/anatomy-physiology.html

McMenamin, P. G., McLachlan, J., Wilson, A., McBride, J. M., Pickering, J., Evans, D. J., & Winkelmann, A. (2018). Do we really need cadavers anymore to learn anatomy in undergraduate medicine? *Medical Teacher*, *40*(10), 1020–1029.

Marković, D., & Marković Živković, B. (2010). Development of anatomical models-chronology. *Acta Medica Medianae*, *49*(2).

Meyer, E. R., James, A. M., & Cui, D. (2018). Hips don't lie: Expert opinions guide the validation of a virtual 3D pelvis model for use in anatomy education and medical training. *HAPS Educator*, *22*(2), 105–118. doi: 10.21692/haps.2018.023

Microsoft HoloLens (2020). Retrieved from https://www.microsoft.com/en-us/hololens

Nicholson, D. T., Chalk, C., Funnell, W. R. J., & Daniel, S. J. (2006). Can virtual reality improve anatomy education? A randomised controlled study of a computer-generated three-dimensional anatomical ear model. *Medical education*, *40*(11), 1081–1087. doi:10.1111/j.1365-2929.2006.02611.x

Nova Southeastern University (2020). I.D.E.A. labs: Virtual reality lab. Retrieved from https://nsufl.libguides.com/idealabs/vr

Oculus (n.d.). Retrieved from https://www.oculus.com/

Pantelidis, P., Chorti, A., Papagiouvanni, I., Paparoidamis, G., Drosos, C., Panagiotakopoulos, T., . . . & Sideris, M. (2018). Virtual and augmented reality in medical education. In *Medical and surgical education-Past, present and future* (pp. 77–97). Retrieved from https://scholar.google.com/scholar?hl=en&as_sdt=0%2C41&q=Virtual+and+Augmented+Reality+in+Medical+Education&btnG=

Penn State (2020). Harrell health sciences library: Virtual reality. Retrieved from https://hershey.libraries.psu.edu/tech-help/library-technologies/technology-sandbox/virtual-reality

Physiology (n.d.). In Merriam-Webster.com dictionary. Retrieved from https://www.merriam-webster.com/dictionary/physiology

Richardson, D. (2011). Is virtual reality a useful tool in the teaching of physiology? *Advances in Physiology Education, 35*, 117–119. doi:10.1152/advan.00002.2011.

Sharecare (2020). Retrieved from https://www.sharecare.com/pages/vr

Singh, V., & Kharb, P. (2013). A paradigm shift from teaching to learning gross anatomy: Meta-analysis of implications for instructional methods. *Journal of the Anatomical Society of India*, *62*(1), 84–89. doi: 10.1016/S0003-2778(13)80019-6

Standring, S. (2016). A brief history of topographical anatomy. *Journal of Anatomy*, *229*(1), 32–62. doi: 10.1111/joa.12473

The Body VR (2020). Retrieved from https://thebodyvr.com/

Virtual Reality & 360 Video Services (2020). University of New Mexico. Retrieved from https://libguides.unm.edu/vr/ap

Visible Body (2020). Retrieved from https://www.visiblebody.com/

University News (2019, April 9). Case Western Reserve, Cleveland clinic launch new era of health education with Sampson Pavilion dedication. Retrieved from https://thedaily.case.edu/case-western-reserve-cleveland-clinic-launch-new-era-of-health-education-with-samson-pavilion-dedication/

Warren, L. A. (2019, March 27). Virtual reality devices arrive as JSU launches Mississippi's first university VR academy. Retrieved from http://www.jsumsnews.com/?p=39565

zSpace (2020). Retrieved from https://zspace.com/

Zweifach, S. M., & Triola, M. M. (2019). Extended reality in medical education: Driving adoption through provider-centered design. *Digital Biomarkers*, *3*(1), 14–21.

Chapter 3

Humanities

Shelly Stone-Moye

Courses in the humanities are a standard component of the educational curriculum in most colleges and universities around the world.

> The term "humanities" includes, but is not limited to, the study of the following: language, both modern and classical; linguistics; literature; history; jurisprudence; philosophy; archaeology; comparative religion; ethics; the history, criticism and theory of the arts; those aspects of the social sciences which have humanistic content and employ humanistic methods. ("About the National Endowment for the Humanities," n.d.)

"For decades humanities scholars have used information technology to analyze the printed word, beginning in the 1940s with a massive concordance project by Roberto Busa, a pioneer in the field" (Frederiksen, 2012, p. 445). "Busa began what even to this day is a monumental task: to make an *index verborum* of all the words in the works of St. Thomas Aquinas and related authors, totaling some 11 million words of medieval Latin" (Schreibman, Siemens, & Unsworth, 2004, para. 3).

Marija Dalbello (2011) researched the genealogy of digital humanities and identified each period of development by a leading authority, technology and infrastructure, artifacts, and documentary form. Today with the advent of personal computers, the Internet of Things (IoT), Artificial Intelligence (AI), machine learning, and cloud computing, humanities and computing have evolved into a new field of study, the digital humanities. "Digital humanities now include other areas of study that go beyond sophisticated textual analysis to include advanced topics in electronic publication, document encoding, new multimedia, digital libraries, augmented reality, interactive gaming, and more" (Frederiksen, 2012, p. 445).

"Virtual Reality (VR), Augmented Reality (AR), and AI [are changing] how we teach and learn. Classrooms of the future will be a hyper-immersive experience that caters to an experiential learning environment through a purely digital ecosystem" (Shenoy, 2020, para. 1). This chapter will focus on present-day virtual integration of digital humanities and the educational applications of advanced technologies used in classrooms and labs in higher education.

DIGITAL HUMANITIES CENTERS

"Digital Humanities Centers (DHC) constitute the most visible incubators and promoters of digital humanities work. Centers may be free-standing, part of a department or their own department, or an arm of support services and thus lodged in the library or in Information Technology (IT)" (Poole, 2017, p. 95). In 2008, Diane M. Zorich, a cultural heritage consultant specializing in planning and managing the delivery of cultural information, conducted a yearlong survey of DHCs in the United States. The project was commissioned by the Council on Library and Information Resources (CLIR).

Zorich (2008) defined DHCs as entities where new media and technologies are used for humanities-based research, teaching, and intellectual engagement and experimentation. The goals of the centers are to encourage humanities scholarship, create new forms of knowledge, and explore technology's impact on humanities-based disciplines. The results of Zorich's survey grouped the functions of DHCs into two general categories:

- Center focused: Centers organized around a physical location, with many diverse projects, programs, and activities undertaken by faculty, researchers, and students. These centers offer a wide array of resources to diverse audiences.
- Resource focused: Centers organized around a primary resource, located in a virtual space that serves a specific group of members. All programs and products flow from the resource, and individual and institutional members help sustain the resource by providing content, labor, or other support services. (Zorich, 2008, p. 1)

The UCL Centre for Digital Humanities (UCLDH) reported in 2011 that there were 144 DHCs around the world, with 44 of them in the United States (Terras, 2011). Most DHCs offer some, or all of the activities listed in Table 2. Between 2007 and 2011, the National Endowment for the Humanities funded

250 projects through its Office of Digital Humanities totaling $15,268,130 (Terras, 2011).

Zorich's (2008) research also concluded that DHCs address the full range of humanities disciplines, while others focus on one or more humanities discipline(s) that form the core of their scholarly or pedagogical pursuits. The research domains of the surveyed centers can be categorized as follows:

> *The humanities (and beyond):* Centers whose research domains encompass all of the humanities (and frequently the interstitial areas between the humanities), the social and natural sciences, the arts, and technology.
>
> *Discipline-specific:* Centers that focus on particular disciplines within the humanities or social sciences, or both, such as history, English, literature, art history, or architecture.
>
> *Humanities pedagogy:* Centers concerned with teaching and instructional methods for learning in the humanities.
>
> *Experimentation:* Centers that explore new methods of creativity or that challenge existing notions about cultural products in a digital arena. (Zorich, 2008, p. 8)

DHCs are the epicenter for VR in humanities. The research and implementation of these various technologies have improved and enriched the teaching and learning experience for researchers, faculty, and students. "Scholars have begun to look to computationally generated 3D models, animations, virtual reality, augmented reality, and games to represent present, past, future, and fictive environments and objects" (Grayburn, Lischer-Katz, Golubiewski-Davis, & Ikeshoji-Orlati, 2019, p. 13).

The advancement of technology makes it possible to change a traditional classroom into an interactive and immersive learning environment. "The release of affordable, fully functional VR headsets in 2016 (most notably, the HTC Vive and Oculus Rift headsets) is fueling experimentation in a variety of academic contexts, from library-based makerspaces and humanities classes to architecture and design programs and law schools" (Grayburn et al., 2019, p. 1).

Duke University's Wired! Lab for Digital Art History & Visual Culture is an example of how institutions can incorporate project-based lab models to explore ways of thinking about visual and material culture through digital technologies. "Formed in 2009 within the Department of Art, Art History and Visual Studies [. . .] the lab started as an experimental class of five faculty and eight students interested in 3D modeling and the reconstruction of historic buildings and environments" (Grayburn et al., 2019, p. 17).

"Wired! is a learning community of faculty, staff, and students. [It] engages visualization methods to prompt new approaches to pedagogy and scholarship in the study and interpretation of the visual arts, architecture, cultural heritage, and built environments" ("About Wired!: History," 2020).

"Current Wired! Lab collaborators include faculty in art and architectural history, archaeology, visual and media studies; the staff director of the departmental Visual Media Center; the university subject-area librarian for visual studies; and a dedicated digital humanities specialist" (Grayburn et al., 2019, p. 17). The Wired! Lab receives support from the Gladys Krieble Delmas Foundation, the National Endowment for the Humanities, the Getty Foundation, and the Kress Foundation, in addition to internal support from Duke University ("About Wired!: History," 2020, para. 3).

MICRO DIGITAL HUMANITIES

Not all institutions have fully established, fully funded DHCs. However, smaller schools and schools without DHCs can use digital humanities. "Digital humanities practices are often understood in terms of significant scale: *big* data, *large* data sets, or digital humanities *centers*. This emphasis leads to the perception that projects cannot be completed without substantial access to financial resources, data, and labor" (Risam & Edwards, 2017, p. 1).

Salem State University in Massachusetts has been able to work around this perception by employing micro digital humanities. "Micro digital humanities validate scholarly output that does not require digital humanities centers, big data, large data sets, and access to high-performance computing. As an intervention in local digital humanities, it places high value on working with available resources, however small" (Risam & Edwards, 2017, p. 1).

> Salem State is the most diverse state university in Massachusetts (35% students of color, 40% first-generation college students) and draws a primarily regional, working-class student population. Both the students and the university have few resources. . . . To serve this population, [they] developed the Digital Scholars Program, piloted through a small grant from the university. [They] designed the program to answer the call of the university's strategic plan to foster student success through research opportunities. Students apply to become Digital Scholars, and those who are selected are mentored through the process of creating a small-scale digital humanities project over the course of a semester. (Risam & Edwards, 2017, p. 1)

Salem State's approach to micro digital humanities is embedded in the university's strategic plan to foster student success through research. The Digital

Scholars Program is student-led and student-centered and only uses freely available resources or open-source software [the university] can host. Using these methods of accessing digital resources is cost-effective for the university and students (Risam & Edwards, 2017).

> Projects students have undertaken include recovering the history of LGBTQ activism at the university, revealing the colonialist gaze of Salem residents who traveled to India in the 1920s, creating historical models of the university's oldest campus building, and connecting contemporary student activism around Black Lives Matter to the history of antiracist activism at the university in the 1970s. (Risam & Edwards, 2017, p. 2)

Risam and Edwards (2017) offer a model for developing digital humanities at scales appropriate to institutional contexts and strategic planning. "Perhaps, more importantly, their model also offers a vision of digital humanities with learning curves and barriers to entry that do not require affiliation with centers, access to expensive technologies, or substantial resources" (Risam & Edwards, 2017, p. 2). Colleges and universities can implement digital technologies in humanities classes in several ways ranging from fully funded DHCs like at Duke University to smaller micro digital humanities projects at Salem State University to individualize classroom instruction and activities.

VR AND AR IN THE HUMANITIES CLASSROOM

DHCs and micro digital humanities projects represent a more structured and institutional and systematic approach to incorporating new and advanced digital technologies in the humanities field. However, there are more pragmatic approaches to using VR tools in various humanities classrooms at colleges and universities that do not have DHCs. "[VR] utilizes interactive 3D images which users can experience with a mouse and a keyboard or with headsets designed to immerse the user in a virtual environment. . . . The lifelike simulations provide authentic experiences for these users" (Pilgrim & Pilgrim, 2016, p. 92).

Barrado-Timón and Hidalgo-Giralt's (2019) research revealed that VR can improve information and knowledge concerning a destination and strengthen the cultural experience of the user by generating new ways of relating to cultural heritage. Cultural exhibitions are created by humanists and used to supplement learning in a wide array of humanities-related fields. VR tools amplify the visual experience. "Viewing and experiencing lifelike simulations complements classroom instruction in many ways. For the classroom teacher, the ability to share experiences relevant to an object or place may provide the background knowledge necessary for a student to gain an understanding of content or text" (Pilgrim & Pilgrim, 2016, p. 92).

> Immersive technologies such as VR can bridge the gap between physical history and cultural heritage data. 3D technologies play a crucial role in providing alternative modes of access to the physical world. Publicly accessible VR systems give users of both university and public libraries an opportunity to experience immersive content. (Grayburn et al., 2019, p. 34)

Ramsey (2017) noted that virtual reality worlds have the potential to communicate and explore the broader historic environment as an experience. The application of these immersive environments allows students the opportunity to explore historical sites, tour museums, analyze ancient artifacts, learn languages, and study famous works of art. "Virtual solutions allow students to forgo expensive travel by practicing their language skills in VR scenarios with AI-powered animated characters. Students can feel less inhibited than in real-life interactions to improve their language skills and receive real-time feedback" (Shenoy, 2020, para. 8).

Grayburn et al. (2019) suggest that the dream of perfect hybridity in teaching with digital technologies is elusive. While the use of some tools, such as readily accessible web-based presentation tools, is easy to teach in a single training session, more involved GIS, 3D, and VR/AR topics require greater degrees of scaffolding and explication.

CHALLENGES FOR COMMUNITY COLLEGES

The National Survey of Digital Humanities in Community Colleges was conducted in 2013. The survey was supported by a National Endowment for the Humanities Office of Digital Humanities Start-Up Grant, entitled Bringing Digital Humanities to the Community College and Vice Versa (National Survey of Digital Humanities in Community Colleges, (2013).

The survey revealed that the top three institutional obstacles to digital humanities at community colleges were a lack of clarity about the difference between digital humanities and other kinds of technology, lack of clarity about learning goals of digital humanities, and a lack of interest in what seems to many to be trendy (National Survey of Digital Humanities in Community Colleges, 2013). The results indicate that community college faculty who teach humanities-related courses can benefit from more training and professional development on how to incorporate digital technologies in the classroom.

In a 2019 training publication titled "Possibilities for Digital Humanities at Community Colleges," Lisa Spiro outlined techniques that community college faculty can use to incorporate digital projects into their curriculum and assist students with developing technical skills and ensuring access to resources. Spiro (2019) suggested that faculty begin the process of incorporating digital projects by determining how digital pedagogy advances course objectives; start small, ideally with content or tools that they are already familiar with, and plan for time required for integrating digital tools and focus on the concepts rather than tools.

Before implementing assignments, Spiro (2019) recommends that faculty try out the assignment first; survey students to understand their skills and needs, as well as what technologies they have access to; use simple, freely available tools; offer step-by-step, hands-on labs, ideally in-class; partner with IT and the library; and recruit student tech mentors.

Spiro's (2019) recommendations address a faculty approach to incorporating digital technologies in the classroom. However, in order for the successful implementation of advanced technologies in humanities classrooms, colleges will have to make an institutional commitment to supporting these technologies. Providing educators and administrators with the necessary resources to effectively carry out the objectives and goals of any successful initiative is vital. Professional development for faculty and staff will be key, as well as the institution's commitment to expanding its IT infrastructure.

In addition to leveraging state and local funding for humanities projects, community colleges should take advantage of federally funded grant opportunities. The National Endowment for the Humanities (NEH) is one of the largest funders of humanities programs in the United States. Each year it makes funding opportunities available to community colleges through its Humanities Initiatives at Community Colleges. This initiative is designed to strengthen the teaching and studying of the humanities at community colleges by developing new humanities programs, resources (including those in digital format), or courses, or by enhancing existing ones. ("About the National Endowment for the Humanities," n.d.)

BENEFITS OF DIGITAL HUMANITIES

Anne B. McGrail, who teaches writing and literature at an open-access community college in Eugene, Oregon, authored chapter 2 of Gold and Klein's book, *Debates in the Digital Humanities 2016*. In the book, she writes, "Many aspects of [digital humanities] seem perfectly fit for the local, open-access missions of community colleges. It is an opportune time to bring [digital humanities] to the community colleges" (Gold & Klein, 2016, p. 17, 18).

Supporting digital technologies in humanities encourages "authentic inquiry; enables students to have a public impact; fosters social learning; promotes creative confidence; and facilitates a greater understanding of the digital environment" (Spiro, 2019, slide 44). The incorporation of these skills in today's technologically advancing world will have a significant impact on student success.

CONCLUSION

Higher education leaders should embrace changes that are being created by the rapid growth of VR and other forms of intelligent technology. Technology trends have impacted all curriculum programs, including humanities, and will continue to do so as technology continues to evolve. "The ability to fully leverage these technologies will enable higher education institutions to transform learning and the campus at large to meet student needs . . . the innovations of VR, and deep learning can improve student success and satisfaction exponentially" (Shenoy, 2020, para. 16).

CHAPTER SUMMARY

- Today with the advent of the personal computers, the Internet of Things (IoT), Artificial Intelligence (AI), machine learning, and cloud computing, humanities and computing have evolved into a new field of study, the digital humanities.
- Zorich (2008) defined DHC as an entity where new media and technologies are used for humanities-based research, teaching, and intellectual engagement and experimentation.
- DHCs are the epicenter for VR in humanities.
- The advancement of technology makes it possible to change a traditional classroom into an interactive and immersive learning environment.
- Duke University's Wired! Lab for Digital Art History & Visual Culture is an example of how institutions can incorporate project-based lab models to explore ways of thinking about visual and material culture through digital technologies.
- Not all institutions will have fully established or fully funded DHCs. However, smaller schools and schools without DHCs can use digital humanities.
- Salem State's approach to micro digital humanities is embedded in the university's strategic plan to foster student success through research.

- However, there are more pragmatic approaches to using VR tools in various humanities classrooms at colleges and universities that do not have DHCs.
- The application of these immersive environments allows students the opportunity to explore historical sites, tour museums, analyze ancient artifacts, learn languages, and study famous works of art.
- Providing educators and administrators with the necessary resources, including professional development, to effectively carry out the objectives and goals of any successful initiative are vital.
- The National Endowment for the Humanities (NEH) is one of the largest funders of humanities programs in the United States.
- Community college leaders should embrace changes that are being created by the rapid growth of VR and other forms of intelligent technology.

REFERENCES

About the National Endowment for the Humanities. (n.d.). Retrieved from https://www.neh.gov/about

Barrado-Timón, D. A., & Hidalgo-Giralt, C. (2019). The historic city, its transmission and perception via augmented reality and virtual reality and the use of the past as a resource for the present: A new era for urban cultural heritage and tourism? *Sustainability, 11*(10) doi: http://dx.doi.org/10.3390/su11102835

Dalbello, M. (2011). A genealogy of digital humanities. *Journal of Documentation, 67*(3), 480–506. https://doi.org/10.1108/00220411111124550

Frederiksen, L. J. (2012). Digital humanities now. *Choice, 50*(3), 445. Retrieved from https://login.proxy200.nclive.org/login?url=https://search.proquest.com/docview/1173367966?accountid=15065

Gold, M. K., & Klein, L. F. (2016). Debates in the digital humanities 2016. Minneapolis, MN; London: University of Minnesota Press. Retrieved from https://dhdebates.gc.cuny.edu/read/untitled/section/14b686b2-bdda-417f-b603-96ae8fbbfd0f

Grayburn, J., Lischer-Katz, Z., Golubiewski-Davis, K., Ikeshoji-Orlati, V., & Council on Library and Information Resources. (2019). 3D/VR in the academic library: Emerging practices and trends. Council on Library and Information Resources.

National Survey of Digital Humanities in Community Colleges. (2013). Retrieved from https://www.surveymonkey.com/results/SM-CL2CVR/

Pilgrim, J. M., & Pilgrim, J. (2016). The use of virtual reality tools in the reading-language arts classroom. *Texas Journal of Literacy Education, 4*(2), 90–97.

Poole, A. H. (2017). The conceptual ecology of digital humanities. *Journal of Documentation, 73*(1), 91–122. doi:http://dx.doi.org/10.1108/JD-05-2016-0065

Ramsey, E. (2017). Virtual wolverhampton: Recreating the historic city in virtual reality. *ArchNet-IJAR: International Journal of Architectural Research, 11*(3), 42–57. doi: http://dx.doi.org/10.26687/archnet-ijar.v11i3.1395

Risam, R., & Edwards, S. (2017, August). Micro dh: Digital humanities at the small scale. In *DH*. Retrieved from https://doi.org/10.4018/978-1-4666-

Schreibman, S., Siemens, R., Unsworth, J. (2004). *A companion to digital humanities.* Oxford: Blackwell, 2004. Retrieved from http://www.digitalhumanities.org/companion/

Shenoy, R. (2020). VR, AR and AI will transform universities. Here's how. Retrieved from https://unbound.upcea.edu/online-2/online-education/vr-ar-and-ai-will-transform-universities-heres-how/

Spiro, L. (2019). Possibilities for digital humanities at community colleges [PowerPoint slides]. Retrieved from https://digitalscholarship.files.wordpress.com/2019/03/digitalpedagogycommunitycolleges2019.pdf

Terras, M. (2011). UCL Centre for Digital Humanities. Infographic: Quantifying Digital Humanities. Retrieved from https://www.ucl.ac.uk/infostudies/melissa-terras/DigitalHumanitiesInfographic.pdf

Wired! (2020). About Wired!: History. Duke University. Retrieved from http://www.dukewired.org/about/.

Zorich, D. (2008). A survey of digital humanities centers in the United States. *Council on Library and Information Resources, 143,* 1–78. Retrieved from http://www.clir.org/wp-content/uploads/sites/6/pub143.pdf

Chapter 4

Criminal Justice

Chris R. Nesbitt

Over the years, those who taught under the umbrella of criminal justice have worked diligently to add value to student learning. Regardless of the institution level—community college, four-year college, or university—instructors and professors have traditionally used guest speakers, mock crime scenes, mock courtrooms, videos, field trips, and internships to improve tactile and experiential learning opportunities for students.

Students learn differently, and various teaching methods keep students engaged and focused, which enhances understanding. Students love hands-on learning. Visiting prisons, juvenile detention centers, and completing an internship add a higher understanding of the material taught in the classroom. Criminal justice programs continue using these same experiential learning strategies today.

Advancements did occur, and programs incorporated simulated training such as firearms training simulators. This type of technology allows a student to experience a realistic training situation through high-resolution screens, cameras, projectors, and a computer system. The scenarios enable the student to react to the event, and the instructor can alter the suspect's behavior based on the student's verbal skills, use of cover or protection, and decision-making. This technology adds value to student learning and the understanding of relevant concepts but does not allow student immersion within the scenario.

The use of virtual reality (VR) in criminal justice programs allows students to experience the learning strategies mentioned earlier without ever leaving the classroom. No more liability waivers to collect or approvals from the program chair and dean to take a field trip. Virtual reality is a three-dimensional, computer-generated environment that allows one to interact in a virtual world and manipulate objects and perform specific tasks (Virtual Reality Society, 2020).

This chapter seeks to share how colleges and universities use virtual reality to improve student learning and prepare students for careers in the criminal justice realm. The majority of virtual reality incorporation lies in the areas of crime scene investigation and forensic science. However, law schools now use virtual reality in classrooms and seek ways to integrate technology with courtroom litigation. Correctional institutions use virtual reality to improve the job interview skills of offenders to aid in higher employability and reduce recidivism.

USES OF VIRTUAL REALITY IN CRIMINAL JUSTICE

A need exists to explain virtual reality in greater detail. Many know the term, but far less understand how it works or the purpose. Virtual Reality Society (2020) explains that virtual reality creates an environment to experience the event as if there. The use or incorporation of virtual reality provides a significant alternative to learning activities that are cost-prohibitive, time-consuming, or too dangerous to complete. The colleges and universities discussed in this chapter understand this logic and made the most of this opportunity.

George Washington University

According to Lieberman (2018), the Police and Security Studies program at George Washington University in Washington, DC, shifted to the virtual world to improve course participation and learning. The program had relied heavily on online and hybrid classes taught asynchronously and synchronously to capture more students. These students could learn from current and former police officers in the area of crime scene investigation. However, the instructors felt like students missed out on the tactile component of instruction. Courses in the program often captured and discussed the missteps made in high-profile investigations like the O. J. Simpson case.

How did this institution incorporate virtual reality into online classes? Program leaders developed a virtual crime scene and a virtual evidence lab to enable students to examine materials collected from the scene. The program staff created iterations of events, which allows students to participate multiple times without duplication. Each event or scenario contains various Easter eggs, which identify areas on which students should focus. Each egg prompts discussion about the investigative process and guides students toward predetermined conclusions (Lieberman, 2018).

What about face-to-face sections? Do the students in these classes have the opportunity to experience crime scene investigation in a virtual world? Lieberman (2018) reports these students can experience simulation via Google Cardboard units purchased explicitly for the project. Students can also access and participate in less immersive versions on smartphones, tablets, or computers. To create the simulation, the program staff took 200 pictures with a 360-degree camera of an artificial but realistic crime scene.

Many individuals worry about the quality or the level of realism of the scenarios using this type of technology. Jeff Delinski, Program Director, had the same concerns. Delinski wondered if the tool possessed the impact to validate its cost. Delinski said, "I was impressed with the level of quality" and he went on to say, "what I like most about it is it can be scaled with added features" (Lieberman, 2018, para 11).

Initially, program staff worked to create a virtual world but needed to begin with a concept for a crime scene. Lieberman (2018) indicated that these individuals wrote a detailed crime story, which included how and where the homicide took place, the evidence's location, and what should result from the investigation. Next, the group took photos of the crime scene, edited the images, and placed them into a platform designed for 360-degree images called ThingLink. Two classes incorporated the use of virtual reality, Crime Scene Investigation and Introduction to Forensic Science. The benchmark for future use includes students developing a working knowledge of the skills each gained from the simulation.

What worked. Lieberman (2018) reports that the simulation with 360-degree images seemed less daunting to create, and it minimized disorientation compared to simulation made from 360-degree videos. To provide a perspective of these photographs, think about how images look on Google Street View. This process allows students to move through the area at one's pace and not get led along by a scripted narrative. Also, the creation allowed for the addition of outdoor sounds to increase the level of realism. Another advantage includes cost-effectiveness. Current costs include:

- Kodak PixPro camera: $500.00
- ThingLink subscription: $150.00
- Adobe Photoshop subscription: $250.00/year
- Product development team salaries (varies)

The simulation worked exceptionally well to springboard discussions that form foundational knowledge and understanding of concepts. Specific questions include:

> When you arrive at a crime scene, what do you look at first? What are you allowed to touch? How do you decide between crucial evidence and red herrings? Where else do you look once you have leads? What questions do you ask in the evidence lab to narrow down your findings? (Lieberman, 2018, para 18)

Most important, anecdotal surveys suggest students and faculty give this experience two thumbs up (Lieberman, 2018).

When one tries something new, rest assured something will go wrong. Lieberman (2018) noted that staff encountered issues with the first camera, Theta. The apps crashed regularly, and they wasted a lot of time. Editing photographs was a time-consuming endeavor. However, using a more sophisticated Adobe tool made editing faster. Also, the staff found it challenging to add crime scene tape around the area and to adequately number the evidence markers.

Center for Innovation through Visualization and Simulation (CIVS)

The CIVS is a collaborative effort between Purdue University Northwest and the University of Illinois at Chicago (UIC). This collaboration resulted in the creation of software that allows students to function in a virtual crime scene (Jacobs, 2018). An advantage of any learning environment is allowing students to make mistakes without fear of repercussions or penalties. "It is a good place to learn. It is a good place to experience a little bit of critique if you did something wrong, compliment if you did something right" (Jacobs, 2018, para 7).

Jacobs (2018) continued to explain that the CIVS allows students to experience hands-on training without contaminating or compromising a crime scene. Larson said:

> When you have a new individual who's learning, the first thing that's going to happen is something is going to go wrong. They're going to forget their gloves, and they're going to leave a fingerprint behind somebody else will find. (Jacobs, 2018, para 9)

Simple errors could lead to extra work for investigators on an active case. A contaminated crime scene opens doors for defense attorneys to question one's competency in performing required duties.

Jacobs (2018) explains how students explore crime scenes in a virtual world at CIVS. The students wear virtual reality headsets and hold controllers, which allow students to interact with the screen. As a student moves around the room in the CIVS, the student simultaneously moves around in the virtual crime scene. The instructors track the student's movements and identify items each one may have missed. For example, when a student placed an evidence marker on the floor, the instructor told the student to look at the footprint they just left on the tiled floor.

How does a student who participates in the virtual crime scene see his or her activities? Jacobs (2018) described how a large TV illustrates each move a student made from the student's perspective. This particular example showed a student who looked around the bedroom and observed a painting missing from the wall. The student also placed a ruler beside a piece of scrap paper and took a screenshot of it. Also, the student used ultraviolet light to look for other evidence he or she might have missed.

According to Jacobs (2018), the software contains approximately a dozen scenarios, which students created over a year. However, an instructor can manipulate each scene in case a student previously completed it. It allows the student to learn something new each time through a scenario. An instructor possesses the capacity to add a piece of evidence, or take evidence away, or hide the evidence. The scenario possibilities are endless.

Returning to the scenario examples, these students do not stop the investigation after identifying and labeling evidence. Jacobs (2018) details how these students collect the evidence and proceed to the virtual crime lab. Wearing protective equipment, gloves, and glasses, students spray the scrap paper collected for possible fingerprints. This process allows students to connect essential variables from the crime scene to the lab. Being able to process evidence in a virtual environment stresses to students the importance of identifying and collecting evidence correctly.

Jacobs (2018) shared what the future holds at the CIVS. The direction of the program emphasizes the development of additional scenarios with a focus on forensic science, specifically, creating scenes that allow students to examine bullet trajectories and blood spatter evidence. Furthermore, the CIVS wants to go public with the virtual crime scene simulations. The CIVS would provide the training for any school or police department interested. Each entity only needs to purchase the software and virtual reality gear to have unlimited access to the virtual-world experience.

Radford University

Radford University, a public university located in Radford, Virginia, reports that the criminal justice program at the university implemented virtual reality into the classroom (Williams, 2019). This effort emphasizes the importance of students gaining realistic experience before entering the profession and making difficult life-and-death decisions. Thus far, students love the technology and some were grinning from ear to ear after the experience. The incorporation of virtual reality in the classroom allowed students to take prison tours without leaving the building.

Williams (2019) indicates program professor Benjamin Wright learned he could do more with the technology. Wright quickly moved to have students enter a virtual crime scene to act as investigators in the famous Unabomber case. Students use the HTC virtual reality headsets, which create an immersive, realistic experience. One student commented that technology makes him feel like he is standing in the real scene. This technology provides students with a better perspective and understanding of the expectations for the profession.

Williams (2019) documents how the criminal justice program at Radford University included virtual reality technology. Wright learned how the State of Colorado began using virtual reality to help offenders reenter society after release. This concept sparked an interest, and he started small. He introduced the technology to a few students with minimal equipment. However, based on the successes of the implementation, Wright plans to incorporate the technique and purchase additional equipment.

Savannah State University (SSU)

The forensic science program at SSU became the first college in Georgia to offer a bachelor of science (BS) degree with tracks in biology and chemistry (Pine, 2019). SSU leads the way by providing students with access to the state-of-the-art technology to enhance one's experience before entering the profession (Pine, 2019). How did SSU develop such a unique learning opportunity for the institution?

Karla-Sue Marriott, PhD, and interim chair of the Forensic Science program, worked with Vizitech to create the world's first virtual reality crime scene software (Pine, 2019). Students wear virtual reality headsets and use hand controllers to select forensic tools and weapons, position bodies, calculate gunshot trajectories and blood spatter, and recreate or predict scene outcomes (Pine, 2019).

Pine (2019) reported that SSU began using the software in 2017 for the program's capstone courses, Crime Scene One and Crime Scene Two. Crime scene investigators now use 3D scanners to document crime scenes. The program supports the use of a FARO 3D scanner to capture and record crime scenes. This technology allows students to take a 3D scan of a room and use virtual reality equipment to enter and explore the area in an immersive capacity. Students gain realistic training to prepare how to skillfully and methodically investigate a crime scene as law enforcement officers.

> The creation of a 3-D interactive and virtual forensics training and simulations tool rich in visual, spatial and kinesthetic content with assessment capabilities offers the ability to provide each student with a first-person, realistic interaction with experiential content. (Pine, 2019, p. 1)

Is it possible for a student to assume the role of a suspect and look at the situation through a different lens? Yes. Pine (2019) explains that the use of a visual avatar enables students to visually and kinesthetically assume the role of a suspect to better understand the physical and emotional possibilities akin to crime scene investigation. Also, with this technology, students could safely conduct experiments such as those related to gunshot projectile trajectories. Virtual technology allowed SSU to shift the forensic science educational paradigm and raise the training bar to new heights.

Incorporating this type of technology into the forensic science program has helped to attract students. Pine (2019) shared that one student toured colleges across the southern United States before making a selection. The student selected SSU because of the university's forensic science facilities. The student said, "being able to recreate crime scenes and solving automated crime scenes has been my favorite part of the program. It helps keep your mind thinking outside of the box" (Pine, 2019, p. 2).

The SSU forensic science program will gain a greater audience soon. SSU decided to begin an 18-hour certificate program in Virtual Forensic Science: Evidence Processing, Documentation, and Analysis. A student in any major can enroll in this certificate program. Moreover, professionals employed in a law enforcement capacity, military, or a relevant area of forensic science, criminal justice, or homeland security can begin the program. The program provides an immersive learning experience using 3D virtual reality technology (Pine, 2019).

Valdosta State University (VSU)

Dr. Bobbie Ticknor, associate professor in criminal justice at VSU, used her software development background to bring virtual reality to VSU. She explained the types of virtual reality equipment the college uses for simulations, classes in which simulations get used, and perceptions from involved students. The interview occurred virtually on May 26, 2020.

Ticknor further explained that the creation of the virtual reality crime lab at VSU resulted from a collaborative effort between her dean, the dean of the library, and the chief information officer (CIO). The dean of the library donated some unused space and an office, and Dr. Ticknor purchased equipment from a $15,000 grant. Starting the virtual lab required the extension of its use to the campus population. This inclusive practice led to a variety of approximately thirty disciplines who used the space for virtual field trips. Dr. Ticknor manages the virtual reality lab and regulates the number of students to 12 at a time.

Since the inception of the virtual lab, she expanded its uses, and currently uses virtual reality technology in two classes. The students in these classes experience five to six simulations. These simulations include prison tours, crime scene investigation, de-escalation, and the proper handling of court evidence. Students benefit tremendously from experiencing realistic situations in a controlled and safe environment (B. Ticknor, personal communication, May 26, 2020). Ticknor said:

> Just think if you were a criminal justice student, and you could see a crime scene before you've actually gone out and had to experience it in real life. We can offer experiential learning opportunities here in the classroom where it's safe and controlled, but our students get that valuable experience just like if they were out in the field. (Umholtz, 2019, p. 1)

She incorporates virtual reality technology into classes immediately. Typically, students begin using the technology at the end of the first week of the semester. During a week of class, she uses two days for lecturing and two days in the lab. This particular schedule seems to work well for the students. It allows for immediate reinforcement of material taught during lecture sessions (B. Ticknor, personal communication, May 26, 2020).

Ticknor provided examples of the different virtual reality levels. First-generation technology requires the use of a cell phone with a virtual reality application to get inserted into a headset like Google Cardboard. The Oculus Go provides an example of second-generation technology. This technology allows one to move around without cords connecting to devices. Third-generation virtual reality equipment includes the Oculus Quest, which enables one to access virtual-world scenarios. Finally, the HTC gives an example of fourth-generation technology, which provides the most realism but is the most expensive.

Ticknor explained that VSU students are not the only individuals who benefit from the virtual reality lab. Current law enforcement officers and law enforcement recruits use virtual reality technology to gain experience with active-shooter scenarios. It allows these individuals to practice similar events before responding to a live situation. Agencies could incorporate virtual reality scenarios during the law enforcement hiring process to aid in the decision-making process for continuing or terminating potential employment opportunities (B. Ticknor, personal communication, May 26, 2020).

Is it possible to use virtual reality technology in correctional reentry programs? Individuals incarcerated for long periods and released from prison into the community have not witnessed the technological changes experienced by the rest of society. Simple everyday practices like self-checkout or mobile ordering may not have existed before incarceration. VSU works with these individuals through the help of virtual reality to foster life skills such as doing laundry, interview skills, computer skills, and self-checkout. Acquiring and developing these skills before getting released dramatically improves one's adaptability to society (B. Ticknor, personal communication, May 26, 2020).

Next steps. Ticknor envisions incorporating virtual reality into online instruction. The associated student cost can get expensive with this addition. However, the purchase of a virtual reality headset can replace the expense of the textbook. The student can use the headset for four years of study, which helps with cost-effective measures. Also, the inclusion of virtual reality instruction to online courses can transform this type of modality and improve student retention and persistence (B. Ticknor in personal communication, May 26, 2020).

EFFECTIVENESS OF VIRTUAL REALITY IMPLEMENTATION IN CRIMINAL JUSTICE PROGRAMS

This chapter illustrated several colleges and universities that incorporated virtual reality into criminal justice programs. The programs focused on the implementation of virtual reality in crime scene investigation and forensic science. Thus far, colleges and universities reported anecdotal evidence to the effectiveness and success of the implementation based on student feedback or other nonscientific methods. Research will determine how the use of virtual reality in the areas of crime scene investigation and forensic science impacts student learning. Mayne and Green (2020) of the University of the West of England, Bristol researched the use of virtual reality in crime scene investigation education in an attempt to find out.

Mayne and Green (2020) recognize that many similar educational programs create fictitious or mock-up crime scenes to aid in student learning. However, many limitations create barriers to the effectiveness of this particular practice. These limitations include high operation costs (academic and technician time), accessibility for differently able students, and the number and types of scenarios offered. Because of these issues, technology provides an impactful alternative to improve student learning. The most frequently used virtual reality technology options require the user to wear a headset capable of producing a 3D image and the use of handheld controllers.

RESEARCH DESIGN AND FINDINGS

Research Methods

Mayne and Green (2020) explained the use of two experimental groups for the research. The first group comprised a group of first-year undergraduate students (UGS) studying forensic science. This cohort of ten students received one traditional learning session in a mock-up or staged crime scene to gain a basic understanding of crime scene investigation. This educational and learning session occurred while supervised. The second group consisted of eight staff and postgraduate researchers (S/PGR). Each participant previously taught or researched in the area of forensic science. The members of this cohort possess advanced knowledge of crime scene investigations.

Mayne and Green (2020) used Oculus Quest 128 Gb headsets for the research. Researchers created the virtual crime scene to replicate a lounge. The environment contained a bookcase, sofa, two chairs, a coffee table, a lamp, and a flat-screen television on a stand. The room contained sufficient lighting for proper illumination. The researchers planted three pieces of evidence in the virtual environment. This evidence included glass shards under the window, a small bloodstain on the carpet between the two chairs, and a knife, which the bookcase partially hid.

It is essential to understand how the participants navigated the environment for this investigation. Mayne and Green (2020) described two ways the cohorts moved around the room. First, participants could walk around the environment. Headset sensors detected their movements, which translated into virtual reality movements. Second, cohort members could opt to use thumb movements on hand controllers. One could use a combination of both control schemes.

Mayne and Green (2020) collected three categories of measurements. These measurements included first, the attainment of learning outcomes (LO), second, a questionnaire, and third, user feedback. The LO came from the institution's course syllabus for first-year forensic science students. Participants received instructions to process the crime scene using the photo tool and evidence markers. All participants had 20 minutes in the virtual reality environment. Upon completion, each member made notes on four criteria directly related to the LO.

Results

Mayne and Green (2020) reported the differences between the number of LO achieved between the two groups. The undergraduate students attained a median score of three out of four outcomes achieved. The staff and postgraduate researchers received a median score of four out of four outcomes achieved. Questionnaire results indicated both groups felt satisfied with the pretest training, intuitiveness of the controls, the quality of immersion from the app, and enjoyed the experience overall. The staff and postgraduate researchers felt it easier to navigate around the crime scene. The students indicated that the app did not induce motion sickness.

The questionnaire results seemed insightful. Mayne and Green (2020) noted most comments centered on the satisfaction with the experience. Some group members suggested extra features to include blood pattern analysis, the ability to move furniture, and rulers for marking evidence for a better experience. As one would imagine, the staff and postgraduate researchers suggested the necessity for additional features.

Mayne and Green (2020) conclude the virtual reality technologies provide an effective an engaging method for imparting crime scene skills. Virtual reality options offer a cost-effective solution to traditional strategies. However, the best choice includes incorporating virtual reality to augment conventional methods of crime scene investigation. Most important, the researchers found value in virtual reality integrated into forensic science education to enhance student satisfaction and the achievement of learning outcomes.

Limitations

Everyone learns through trial and error. Situations, circumstances, and experiences deserve reflection to modify practices for future improvement. This research is no different. "Full immersion in the experience will always be limited by both software and hardware; hence, we do not suggest that VR should or will ever fully replace conventional practical-based learning" (Mayne and Green, 2020, p. 11). Furthermore, the initial cost of virtual reality hardware is significant, as well as the cost of development (Mayne and Green, 2020).

OTHER USES FOR VIRTUAL REALITY IN CRIMINAL JUSTICE

Crime scene and forensic science programs are not the only ones implementing virtual reality to help with learning. Law schools and adult corrections researchers use virtual reality in specific ways to improve outcomes and performance. This section looks at each of these areas and how each incorporated virtual reality into a transformational learning paradigm. This technological shift can excel with continued successes.

Law Schools

Persky (2019) reports that the North Texas Dallas College of Law created mock crime scenes and transformed them into a virtual world to aid criminal law classes. Jennifer Wondracek, the director of educational technology at the college, believes virtual reality technology can significantly improve the law school experience.

> We are at the beginning of what I think is going to be a revolution in the way we train our students. We are right at the edge with virtual reality technology in both the law school setting and the legal profession. (Persky, 2019, p. 1)

However, virtual reality technology fails as a fan favorite at most law schools. Persky (2019) describes some reasons for this reluctance. Law school administrators and faculty operate on a limited budget. Due to financial restrictions, these individuals tend to fund other projects. Also, law schools do not tend to run on cutting-edge technology. These institutions wait until technology advances before pursuing it.

A counterargument to this claim focuses on the affordability of virtual reality technology. Persky (2019) explains that $1,000 can provide students with an experiential learning opportunity in a virtual world. As more law schools move to online instruction, this technology would provide an invaluable learning experience for students. The University of Missouri-Kansas City School of Law also uses virtual reality to record students conducting voir dire. It provides a critical way for students and faculty to critique performance. Currently, it serves as an assessment tool. However, the goal is to add virtual reality content to the curriculum, such as virtual courtrooms (Persky, 2019).

Adult Corrections

Matthew Smith of the University of Michigan School of Social Work leads research focusing on improving employment and reducing recidivism among prison offenders ("Improving Employment," 2020). Nearly 600,000 state and federal ex-offenders reenter communities on an annual basis. Approximately 50% repeat criminal activity within the first year of release ("Improving Employment," 2020). Traditional programs exist to help with this cause. However, this research uses virtual reality to aid in job interview training to improve employability and reducing recidivism rates ("Improving Employment," 2020).

Laboratory results indicate positive results from the implementation of the virtual reality job interview tool. These results are based on five randomized control trials to improve job interview skills and employment rates among those with a history of mental illness, substance abuse, and educational disabilities. This tool has yet to get evaluated among ex-offenders but may become an essential innovation in this field. The main focal area includes increases in employment and a reduction in recidivism ("Improving Employment," 2020).

RECOMMENDATIONS FOR VIRTUAL REALITY IMPLEMENTATION

As discussed throughout this chapter, implementation of virtual reality in criminal justice programs in higher education seems sparse. However, those who leaped in gained an immediate return on the investment. Students gained realistic training and experience, enjoyed the opportunity, and had opportunities that they could not experience otherwise. Why do these programs settle for traditional strategies rather than stretching the limits with virtual reality implementation? Some may feel the technology is too cost-prohibitive. Others might seem ill-prepared to implement this type of technology into the curriculum.

Getting started with some virtual-world component to improve classroom learning is the first step. Google Cardboard provides an inexpensive opportunity to experience virtual reality education. The cost of the product is less than $20.00 per set. One needs a mobile device, which contains a virtual reality application to enjoy an immersive experience. With this product, virtual tours of prisons, courtrooms, and crime labs are possible.

Purchasing a 360-degree 3D camera allows one to photograph a mock crime scene. An Adobe application can allow for editing the photographs. Additionally, one would need an app to transform the pictures into a virtual world. One needs a newer generation of virtual reality headsets for this type of immersive opportunity. This method might cost more, but students experience the moment with greater realism.

Implementing virtual reality in a criminal justice program might require a more inclusive approach from the outset. Sharing the technology with other programs across campus can help to initiate the procurement of the necessary equipment. As well, one can apply for grant funding to help offset the initial investment to purchase equipment. Starting small can allow a more significant portion of the college to see the need for continued investment in virtual reality technology.

CONCLUSION

College and university administrators and faculty need to open an ear to virtual reality technology opportunities in criminal justice programs. This chapter highlights several programs that ventured into virtual reality implementation. While the technology remains new to higher education, the trend continues to shift the paradigm in virtual reality for experiential learning. Several reasons exist, and each one provides a foundational framework for integration in higher education.

The affordability and flexibility of virtual reality technology offer various options to programs based on available institutional funding. Available software applications transition those photographs into a virtual world using a virtual reality headset. This method is relatively inexpensive. Research suggests that implementing virtual reality into crime scene and forensic science programs improves student learning and course outcomes. A better understanding of program and practitioner concepts helps students perform better when placed in real-life situations. Virtual reality provides an immersive environment, making the students feel like they arrived at a crime scene.

Safety is also an important variable when considering the implementation of virtual reality technology. This type of technology allows students to explore prisons or navigate infamous crime scenes without leaving the classroom. No superior approval or liability waivers needed. There is no need to leave campus. The virtual reality component provides methods for analyzing blood spatter and bullet trajectories in a safe and controlled environment. These techniques would otherwise seem impossible in a college classroom.

Virtual reality technology can serve as an impactful recruiting tool. Students possessing a keen interest in crime scene investigation or forensic science get drawn to this type of technology. As described earlier, one student selected Savannah State University because of the institution's virtual reality lab. Traditional students today experienced every level of education with some form of technology. These students have a significant comfort level with technology and embrace different uses to improve learning and gain invaluable experience.

CHAPTER SUMMARY

- Students love hands-on learning.
- The use of virtual reality in criminal justice programs allows students to experience the learning strategies mentioned earlier without ever leaving the classroom.

- The majority of virtual reality incorporation lies in the areas of crime scene investigation and forensic science.
- The use or incorporation of virtual reality provides a significant alternative to learning activities that are cost-prohibitive, time-consuming, or too dangerous to complete.
- The Police and Security Studies program at George Washington University in Washington, DC, shifted to the virtual world to improve course participation and learning.
- Lieberman (2018) reports creating the simulation with 360-degree images seemed less daunting to create, and it minimized disorientation compared to simulation made from 360-degree videos.
- The CIVS is a collaborative effort between Purdue University Northwest and the University of Illinois at Chicago (UIC). This collaboration resulted in the creation of software that allows students to function in a virtual crime scene (Jacobs, 2018).
- Jacobs (2018) shared what the future holds at the CIVS. The direction of the program emphasizes the development of additional scenarios with a focus on forensic science.
- Radford University reports that the criminal justice program at the university implemented virtual reality in the classroom.
- Students use the HTC Vive virtual reality headset, which creates an immersive, realistic experience.
- The forensic science program at Savannah State University (SSU) became the first college in Georgia to offer a bachelor of science (BS) degree with tracks in biology and chemistry (Pine, 2019).
- Students wear virtual reality headsets and use hand controllers to select forensic tools and weapons, position bodies, calculate gunshot trajectories and blood spatter, and recreate or predict scene outcomes (Pine, 2019).
- Virtual technology allowed SSU to shift the forensic science educational paradigm and raise the training bar to new heights.
- Dr. Bobbie Ticknor, associate professor in criminal justice at Valdosta State University (VSU), used her background in software development to bring virtual reality to VSU.
- Simulations include prison tours, crime scene investigation, de-escalation, and the proper handling of court evidence.
- Current law enforcement officers and law enforcement recruits use virtual reality technology to gain experience with active shooter scenarios.
- Ticknor envisions incorporating virtual reality into online instruction.
- Research will determine how virtual reality in the areas of crime scene investigation and forensic science impacts student learning.

- Mayne and Green (2020) conclude the virtual reality technologies provide an effective an engaging method for imparting crime scene skills.
- The initial cost of virtual reality hardware is significant, as well as the cost of development (Mayne and Green, 2020).
- Persky (2019) reports that the North Texas Dallas College of Law created mock crime scenes and transformed them into a virtual world to aid criminal law classes.
- Law schools do not tend to run on cutting-edge technology. These institutions wait until technology advances before pursuing it.
- As more law schools move to online instruction, this technology would provide an invaluable learning experience for students.
- As discussed throughout this chapter, virtual-reality implementation into criminal justice programs in higher education seems sparse.
- Getting started with some virtual world component to improve classroom learning is the first step. Google Cardboard provides an inexpensive opportunity to experience virtual reality education.
- College and university administrators and faculty need to open an ear to virtual reality technology opportunities in criminal justice programs.
- Virtual reality provides an immersive environment, making the students feel like they arrived at a crime scene.
- Students possessing a keen interest in crime scene investigation or forensic science are drawn to this type of technology.

REFERENCES

Jacobs, B. (2018, July 3). Virtual crime scene gives students hands-on training without the risk. *Chicago Tribune.* Retrieved from https://www.chicagotribune.com/suburbs/post-tribune/ct-ptb-virtual-crime-scene-st-704-story.html

Lieberman, M. (2018, May 16). Trial and error: Virtual reality "as memorable as the O. J. Simpson case." *Inside Higher Ed.* Retrieved from https://www.insidehighered.com/digital-learning/article/2018/05/16/crime-scene-simulation-uses-virtual-reality-unite-disparate

Mayne, R., & Green, H. (2020). Virtual reality for teaching and learning in crime scene investigation. Manuscript submitted for publication. Retrieved from https://www.preprints.org/manuscript/202004.0434/v1

Persky, A. (2019, January). How are law schools using virtual reality tools in classrooms? *ABA Journal.* Retrieved from https://www.abajournal.com/magazine/article/virtual_reality_augments_law_school_curricula

Pine, A. (2019, September 9). Virtual forensic science: SSU. *Savannah State University.* Retrieved from https://www.savannahstate.edu/News/2019/07/05/Virtual-Forensic-Science-SSU

Umholtz, K. (2019, August 27). Bringing VR to VSU: Students experience virtual classrooms. *Valdosta Daily Times.* Retrieved from https://www.valdostadailytimes.com/news/local_news/bringing-vr-to-vsu-students-experience-virtual-classrooms/article_f941a729-d7ba-5013-84ea-a4a0dfa97914.html

The University of Michigan. (2020). Improving employment and reducing recidivism among prison offenders via virtual reality job-interview training. Retrieved from https://ssw.umich.edu/offices/research/projects-grants/improving-employment-and-reducing-recidivism-among-prison-offenders-via-virtual-reality-job-interview-training/WEB15152

Virtual Reality Society. (2020). What is virtual reality? Retrieved from https://www.vrs.org.uk/virtual-reality/what-is-virtual-reality.html

Williams, Brooke. (2019, February 7). Radford students get hands-on experience with virtual reality. *WFXR Fox News.* Retrieved from https://www.wfxrtv.com/news/local-news/radford-students-get-hands-on-experience-with-virtual-reality/

Chapter 5

Technical Programs

Kenneth N. Buckey

TOMORROW'S TECHNOLOGY TODAY

Remember sitting in the living room and watching *Star Trek* on the television and seeing the characters go to the Holodeck to relax and take part in a virtual simulation of some sort? *Star Trek* was only science fiction, and there was nothing like it except in television and the minds of the show's creators. Fast-forward just a few years, and today that concept is not only here but available on a predictable mass-market format that ranges from virtual reality (VR) headsets to entire VR rooms and gaming consoles. Where does this come into play for higher education?

Virtual reality is a practical and efficient hands-on instruction that is safe, cost-effective, and invaluable to student and instructor teaching and learning. Virtual reality instruction is transforming instruction and learning. Using tomorrow's technology as a means to enhance classroom techniques, higher education institutions are closing the gap between idea and reality. Colleges need to be able to train and prepare students for the future workforce, which includes having qualified instructors and professors who are proficient in VR technology and be able to teach using this technology in their courses.

As leaders in higher education institutions are always looking for ways to improve the learning environment, VR has added an innovative capability with immersive learning. "Immersive learning places individuals in an interactive learning environment, either physically or virtually, to replicate possible scenarios or to teach particular skills or techniques. Simulations, role play, and virtual learning environments can be considered immersive learning" (NA, 2020, para 1).

VR can assist students' learning because it is free of distractions and allows the student to focus on the task at hand. "When virtual reality tools were first introduced into the classroom, the technology was marketed as a way to go beyond the walls of a school and take students places they would otherwise not be able to go" (Zimmerman, 2019, para 11).

With online courses becoming prevalent in today's world and global competitiveness, technological advances, and the need for skilled workers, the responsibility of community colleges and universities to supply the workforce is increasing. The trades of today now require not only smart, learned individuals but those who have some practical hands-on knowledge of the systems and practices related to their field of study. To meet those needs, virtual reality education is just what the doctor ordered.

From personal computers to smartphones, the exponential growth of technologies is moving faster than the speed of light. The internet of things has permeated every aspect of our lives today, including business and industry. Technology and their innovations no doubt shape our lives. It is the educator's job to ensure that the real-time demands for access and accessibility are there for the students. From self-driving cars, quantum computers, Bitcoin currency, artificial intelligence, and virtual reality, the future is ever moving and attainable.

EXISTING PROGRAMS

Several community colleges today offer courses either through the institution's continuing education or curriculum programs that use virtual reality methods. Training programs that provide virtual reality and hands-on practices are better equipped to train the next level of professionals. Not only are VR programs cost beneficial and economically practical, but they also provide feedback and comprehensive evaluation tools for all students and businesses that use them.

Virtual reality programs vary from college to college, and it depends on the area's needs and programs they offer. Each institution of higher education has unique programs that are synonymous to their needs. The question is, which program gets VR to support the workforce needs of industry? The next sections will cover a few program areas where VR is being used in higher education.

WELDING

Welding offers unique challenges both from the safety standpoint and because of the cost of materials; the program also keeps students more productive because they want to use the machines more. Now that virtual reality has come on the scene, students can train anytime as long as needed, the mistakes they make can be erased and redone with on-the-spot corrections and retraining can be had without the danger and extra costs associated with material for practice. Virtual reality opens up career opportunities and engages students on a personal competency goal, which is exciting and fun, as students can keep track digitally of their assignments and practice work.

Due to liabilities and safety concerns, the institution cannot bring prospective students into the actual welding shops. This changes with the use of a simulator. During recruitment activities, the institution can generate interest in the program. From the backyard welder to the professional steelworker, attention can be fostered through the use of virtual reality. As far as a recruiting tool, the machine can be taken to open houses, high school career fairs, and community events to garner interest not only in the program, but a great career opportunity.

Virtual reality training will most likely never totally replace hands-on training; however, virtual reality is a new tool that can blend training opportunities and allow for better engagement of students that enhances not only training but student comprehension and skills.

AUTO PAINTING

Many community colleges, in keeping with their technical role, have collision repair and auto finishing programs as important programs. Training for auto painting or refinishing is much like welding in that requires expensive material and practice time to be proficient at it. With virtual reality, educational institutions can improve their training programs and train skilled and proficient auto painters as well as recruit new students to the program. This type of training is a stepping stone and allows for students to learn the skills and techniques in VR before going on to the real sprayer and painting an actual car. Using VR saves time, the effort to sand and prepare a vehicle before each painting, and the cost of paint.

EMERGENCY MANAGEMENT

Emergency management is a field that often relies on hands-on experiences and lessons learned through case studies in the classroom environment. With the advances in virtual reality, practitioners of emergency management can program lessons learned and case studies into a virtual reality world. Students will have to develop plans to mitigate and respond to natural and human-made disasters in a virtual setting.

One question from the field has been, will this type of training make response training more effective? The answer is yes. Public safety organizations and emergency responders all want realistic training that puts the students in as close to real-world events as possible with safety in mind. The problem has always been that with emergency response training, there is a viable danger to this type of training; so, safety is paramount. Virtual reality allows training in an environment where safety constraints can come off, and allow for mistakes at no cost to life or property. "I believe we can equip and understand more about what a first responder needs with this type of technology; I think Virtual Reality and real-time data processing [are] only meant to enhance current functioning" (Corrigan, 2019, para. 15).

Fayetteville Technical Community College (FTCC) in North Carolina is using a system for training emergency management students that allows for an immersive environment that is capable of producing multiple scenarios from fires to riots. Students have the capability to not only direct the scene but act as members of an operations center without the number of participants and equipment needed for a real-world, hands-on scenario.

The system being used is named the Advanced Disaster Management System, and it "offers challenging, true-to-life virtual environments for training incident command and disaster management teams at all levels. Trainees gain the confidence, practical experience, and decision-making skills needed to resolve real-life incidents successfully" (Environmental Tectonics Corporation, 2020, para 1). The result is responders and students alike gain valuable insight and a higher level of preparedness through this type of training, which in the end increases mitigation programs and decreases damages and loss of life.

HEATING, VENTILATION, AND AIR-CONDITIONING

One program that can benefit from virtual reality is heating, ventilation, and air-conditioning (HVAC) repair courses. "Virtual reality is increasing its role in the HVAC industry," said Amanda Comunale, director of construction piping services, Victaulic Co. "In preconstruction, it's allowing project teams to visualize the job site in new ways" (Kostora, 2017, para 7).

The use of virtual reality in training courses offers several benefits to the student and the institution. In the foundational courses, it allows for hands-on training into the foundational concepts and diagnostics without having to use actual equipment. Students can practice using the proper tools and gaining real-life experience before they even touch the real machines, making them better prepared and confident in their skills. A unique aspect of virtual reality in the field of HVAC is that a problem can be programmed into the simulator and worked on before a technician even gets to conduct a diagnostic test on an actual machine.

AVIATION TRAINING

In the 1930s, the first flight simulator was used to train pilots, with paddles, wires, springs, and enclosed cockpits. Today virtual reality available to train new pilots and keep trained ones current. It is a game changer. "VR provides a synthetic space, which immerses pilots in an incredibly realistic training scenario" (Naveen, 2018, para. 3). Using VR reduces the costs and space required to house a full-size simulator for training and allows more students to train simultaneously.

VR in the aviation industry is increasing, and colleges and universities that offer aviation programs are a great place to showcase this tool and its possibilities. One higher education institute synonymous with the aviation industry is Embry-Riddle Aeronautical University.

> The mission of the VR lab at this institution is to explore, develop, and test immersive simulation technologies for use in aviation and aerospace research learning. Potential research topics include pre/post-flight operations, checklist memorization, three-dimensional navigation, airfield safety, cognitive process, and transfer of training." (Embry-Riddle, 2020, para 1)

Another university using VR is the University of North Dakota's aviation program. "While wearing a headset such as the HTC Vive units, users have a full 360-degree view of their simulated environment. Specifically, in the lab, users can see the full cockpit of a Piper Archer single-engine aircraft exactly as it looks a few miles away at UND's Aerospace Flight Operations, instrument gauges, and all. This capability is best applied to students who are just starting in their flight courses, establishing situational awareness and the basics of stick and rudder skills" (UND Today, 2020, para 4).

Virtual reality is an excellent tool in the aviation industry and training institutes for both pilots and ground maintenance crews. Inspections and walk-around checks are a vital component of aircraft safety. Imagine the ability to conduct these types of tests in a controlled training environment where the instructor can interject many standard systems and structural problems that can be found. "Without necessary skills and knowledge, the ground crew put both pilots and passengers into danger since this is the crew that makes sure that everything's alright with an airplane and it's ready to take off" (NA, n.d., para. 6).

The aviation industry needs qualified and competent engineers to facilitate maintenance, design, and engineering of today's and tomorrow's aircraft. Sound practical knowledge and practical experience of these systems and components require hands-on training with sometimes large pieces of equipment. Virtual reality programs allow students to learn on complex aircraft systems without having actually to have them on-site. This independent time offers new opportunities for colleges and training institutes to give students more time in actual real-world environments without needing the expensive, sophisticated equipment.

> Pratt & Whitney, an American manufacturer of jet engines, uses VR to train its mechanics to repair aircraft engines. With their virtual reality experience, engineers can look inside an engine or observe it around. Also, they can study nearly any particular detail individually using virtual reality. (Bellamy III, 2017, para. 3)

This unique training environment allows students to work in hazardous environments and safely see the workings and problems with an engine. In the classroom environment, using VR headsets allows for a virtual workspace and on-the-spot critiques from the instructor. Using VR in training, an institution will be able to properly train students and make sure they are ready for a variety of situations.

ENGINEERING PROGRAMS

3D printing has been used to develop prototypes and allows for a visual representation of a design, but materials, time, and equipment are needed, and what if the prototype is not correct? Enter VR for engineering, an immersive environment where the student developer can see all sides and all aspects of the design before it is produced. Engineering requires critical thinking, along with mechanical and engineering knowledge. VR allows for an inclusive environment in which students can build, tear apart, and rebuild a system, structure, or component safely and cost-effectively without danger, costly mistakes, and excessive material (Green, 2015).

The College of Engineering at the University of Michigan at Ann Arbor has a lab facility with its structural engineering students work using VR. Students use 3D modeling to build projects and, in turn, gain valuable insight about structurally sound buildings. The lab also provided the entire university with access to new technologies and subject matter experts if their programs would like assistance with VR capabilities.

Computer modeling and visualization of multidimensional environments apply in nearly any discipline. Some of the capabilities of this program area allow for advanced concepts to be visualized and then created with 3D modeling. The range of engineering programs that can use this technology includes mechanical, electrical, and chemical engineering programs.

EXPANSION INTO EVERYDAY PROGRAMS

How students learn has always been a factor in education and a concern for educators at all levels. Most technologies used in training are help aids and information access tools, from e-books to online testing sites for students. Virtual reality is about being there and doing it. Imagine reliving the history of the O.K. Corral and being in the streets with Doc Holliday versus just reading about it or seeing a PowerPoint. Albert Einstein once said, "I never teach my pupils; I only attempt to provide the conditions in which they can learn" (Einstein, n.d.).

Virtual reality (VR) has applications in criminal justice, healthcare, fire protection, emergency management, and many other programs today.

> To help students grasp concepts in Earth science, an Eastern Michigan University professor built an AR sandbox using a Microsoft Xbox Kinect camera, digital data projector, computer, and simulation and visualization software. Students can create mountains, volcanoes, river channels, glacial deposits, or virtual rain by manipulating a digital map projected on a box of sand. (Brereton, 2018, para. 3)

Virtual reality is an excellent tool for recruitment efforts and can give new students and parents a virtual tour of the campus. As an example of this type of recruitment tool, Savanah College of Art and design used VR and sent Cardboard headsets to be used with smartphones to over 30,000 prospective students and, in doing so, had a 12% jump in admissions.

FUTURE IMPLICATIONS

Where is VR going is not the real question, but where has it gone so far? "The VR industry is expected to balloon to a behemoth-sized $120 billion market. Bloomberg goes on to estimate that by 2025 that number will further increase to a whopping $182 billion in size" (Adams, 2016, para 8).

Higher education today is the tip of the spear when it comes to virtual reality platform training. Higher education institutes have opened independent labs on campuses for virtual reality research and immersive learning innovations. Virtual reality is changing quickly, and higher education needs to be well informed of the technology trends and industry's needs for students to benefit from this technology now.

The concepts of VR may seem scary and expensive; however, there are several low-cost resources available to teachers today and readily available. For a geography class, one could use 360cities, an application that allows students to visit any city and get a 360-degree tour.

> Virtual Reality has the power to educate people in a way that we have not been able to before. Immersive education has the potential to break down the huge barriers that traditional learning has always struggled with, through experiential learning. (Immersive Education, 2020, para 1)

VR TUTORING

A unique concept in New York is virtual reality tutoring for K-12 students. This program was started by two Cornell University students and grew into a company that distributes Oculus VR headsets to the student enrolled in the tutoring program.

The future potential for VR is exciting. While science and medical-related programs are engaging in VR now, the integration of virtual reality into all learning environments will have a significant impact soon in higher education. An article by the financial policy council stated, "VR in education is predicted to be a $200 million industry by 2020, and a $700 million industry by 2025" (Silverman, 2019, para 8).

DOWNSIDE

Most any new technology requires extensive cost. Virtual reality costs can be controlled with the use of simple Cardboard viewers connected a cell phone for some rudimentary VR; however, complicated applications and systems will require expensive headsets to operate. Who pays for this? Is this part of the student package much like a book fee or lab fee, will financial aid pay for these extras, or will the community college or university have to come up with the funding?

Another downside to consider is the loss of the social context of universities and colleges and the dynamics campuses provide with integration and diversity. In a virtual world, that person-to-person contact goes away, and that enrichment is lost. The physical connection between a teacher and their students is also lost and replaced with an avatar.

The functionality of using virtual reality software could limit some interactions in class and make the course a more fixed environment. Another disadvantage may be that virtual reality is somewhat addictive, as the user's experience is not a genuine reality, but may become one to them, which could cause disassociation with the outside world.

CONCLUSION

Virtual reality is no longer a science fiction concept or far-fetched idea; it is here to stay. Technology has driven advances in every facet of higher education, from computers to advanced hardware in 3D technologies. The virtual reality environment has some great benefits and uses within higher education and has some downsides as well. Leaders should take the pros and cons into consideration when developing a program with VR.

With global awareness and collaboration, a need for a virtual meeting or common place is expanding the learning environments in which educators will teach. Face-to-face classrooms are being replaced with virtual environments. Virtual reality and its implementation in the classroom gives a positive outcome to students and enhances their engagement in an immersive learning environment that offers real-world possibilities. Virtual reality expectations are increasing every day, and the growth expected over the next few years is exponential.

Institutions that are already implementing virtual reality programs are leading the pack, and those that are not are already behind the times. The future is here today, and tomorrow is only an idea away. Virtual reality in education is a change agent that is going to be an innovative tool for future generations of students. The technology is here, and it is up to educators to adopt and use VR inside their respective classrooms and labs with the mindset of student growth, engagement, and accessibility to the world of global education.

CHAPTER SUMMARY

- Virtual reality is a practical and efficient hands-on method of instruction that is safe, cost-effective, and invaluable to student and instructor teaching and learning.
- VR can assist students' learning because it is free of distractions and allows the student to focus on the task at hand.
- The trades of today now require not only smart, learned individuals but those that have some practical hands-on knowledge of the systems and practices related to their field of study.
- Training programs that provide virtual reality and hands-on practices are better equipped to train the next level of professionals.
- Welding offers unique challenges both from the safety standpoint and the cost of material, the program also keeps students more productive because they want to use the machines more.
- From the backyard welder to the professional steelworker, attention can be fostered through the use of virtual reality.
- Training for auto painting or refinishing is much like welding in that requires expensive material and practice time to be proficient at it.
- With the advances in virtual reality, practitioners of emergency management can program lessons learned and case studies into a virtual reality world.
- Virtual reality allows training in an environment where safety constraints can come off, and allow for mistakes at no cost to life or property.
- Another program that can benefit from virtual reality is heating, ventilation, and air-conditioning (HVAC) repair courses.
- A unique aspect of virtual reality in the field of HVAC is that a problem can be programmed into the simulator and worked on before a technician even gets to conduct a diagnostic test on an actual machine.
- In the 1930s, the first flight simulator was used to train pilots, with paddles, wires, springs, and enclosed cockpits. Today virtual reality is available to train new pilots and keep trained ones current.
- Virtual reality is an excellent tool in the aviation industry and training institutes for both pilots and ground maintenance crews.
- Virtual reality programs allow students to learn on complex aircraft systems without having actually to have them on-site.
- VR allows for an inclusive environment that students can build, tear apart, and rebuild a system, structure, or component safely and cost-effectively without danger, costly mistakes, and excessive material.
- Virtual reality (VR) have applications in criminal justice, healthcare, fire protection, emergency management, and many other programs today.

- Higher education today is the tip of the spear when it comes to virtual reality platform training.
- While science and medical-related programs are engaging in VR now, the integration of virtual reality into all learning environments will have a significant impact soon in higher education.
- Another disadvantage may be that virtual reality is somewhat addictive, as the user's experience is not a genuine reality, but may become one to them which could cause disassociation with the outside world.
- Virtual reality expectations are increasing every day, and the growth expected over the next few years is exponential.

REFERENCES

Adams, R. (2016, March 21). Five reasons why virtual reality is a game-changer. Retrieved May 12, 2020, from https://www.forbes.com/sites/robertadams/2016/03/21/5-reasons-why-virtual-reality-is-a-game-changer/#59e10ae41bed

Bellamy III, W. (2017, August 24). 9 companies using augmented and virtual reality in aviation. *Aviation Today*. Retrieved June 25, 2020, from https://www.aviationtoday.com/2017/08/24/9-companies-using-augmented-virtual-reality-aviation/

Brereton, E. (2018, May 9). How college VR is transforming teaching and learning. Retrieved May 15, 2020, from https://edtechmagazine.com/higher/article/2018/05/4-ways-colleges-are-embracing-virtual-reality

Corrigan, J. (2019, March 6). Virtual reality could make disaster response training more effective. Retrieved May 20, 2020, from https://www.nextgov.com/emerging-tech/2019/03/virtual-reality-could-make-disaster-response-training-more-effective/155357/

Covey, S. M., & Merrill, R. R. (2018). *The speed of trust: The one thing that changes everything.* New York: Free Press. Retrieved April 28, 2020, from https://www.amazon.com/SPEED-Trust-Thing-Changes-Everything/dp/074329730X#reader_074329730X

Einstein, A. (n.d.). Albert Einstein quotes. Retrieved June 3, 2020, from https://www.goodreads.com/quotes/253933-i-never-teach-my-pupils-i-only-attempt-to-provide

Embry-Riddle. (2020). Virtual-reality lab. Retrieved June 25, 2020, from https://daytonabeach.erau.edu/about/labs/virtual-reality

Environmental Tectonics Corporation. (2020). Advanced disaster management simulator. Retrieved June 1, 2020, from http://www.trainingfordisastermanagement.com/

Green, E. (2015, November 16). Using virtual reality to teach tomorrow's engineers. Retrieved June 26, 2020, from https://www.engineering.com/Education/EducationArticles/ArticleID/10990/Using-Virtual-Reality-to-Teach-Tomorrows-Engineers.aspx

Immersive Education. (2020). About us. Retrieved June 1, 2020, from https://immersivevreducation.com/about-us/

Kostora, N. (2017, April 10). Virtual reality's growing role in the HVAC industry. Retrieved June 10, 2020, from https://www.achrnews.com/articles/134822-virtual-realitys-growing-role-in-the-hvac-industry

NA. (2020). Immersive learning. Retrieved May 12, 2020, from https://trainingindustry.com/glossary/immersive-learning/

NA. (n.d.). Augmented reality in aviation training. Retrieved June 25, 2020, from https://jasoren.com/how-virtual-and-augmented-reality-are-used-in-aviation-training-and-other-use-cases/

Naveen, J. (2018, July 23). VR is capable of transforming the aviation industry. Retrieved June 25, 2020, from https://www.allerin.com/blog/vr-is-capable-of-transforming-the-aviation-industry

Silverman, S. (2019, October 14). The future of virtual reality in education. Retrieved June 3, 2020, from https://www.financialpolicycouncil.org/blog/the-future-of-virtual-reality-in-education/

UND Today. (2020, February 3). Simulators on campus: UND Aerospace launches VR flight trainer. Retrieved June 25, 2020, from https://ndus.edu/2020/02/03/simulators-on-campus-und-aerospace-launches-vr-flight-trainer/

Zimmerman, E. (2019, August 19). AR/VR in K–12: Schools Use Immersive Technology for Assistive Learning. *EdTech*. Retrieved May 12, 2020, from https://edtechmagazine.com/k12/article/2019/08/arvr-k-12-schools-use-immersive-technology-assistive-learning-perfcon

Chapter 6

Recruitment

Jeffrey Merritt

RECRUITMENT AT THE COLLEGE LEVEL

College recruitment is becoming competitive between institutions, causing many colleges and universities to be innovative when recruiting in-state and out-of-state students. While it is recommended that every potential student takes advantage of an on-site campus visit, it is worth considering that many interested students may not have the means to visit a campus. Colleges and students alike are now trying to adapt to a new normal: virtual tours are replacing in-person tours, and information sessions are held over Zoom. Potential students will likely have to make a difficult choice about the next four years of their education without ever setting foot on a campus or meeting a representative in person.

This new normal will force colleges to get more creative with their recruiting processes. One option for colleges to reach prospective students is by improving their online visibility and making it easy and fun for prospective students to explore the campus in-depth. Virtual reality (VR) tours allow prospective students to maneuver through the campus without having to physically be there. Using VR-based technologies, colleges can help students get their early orientations of the campus through virtual tours.

Thanks to 360-degree campus virtual reality (VR) tours, a student from anywhere in the world can visit several foreign campuses in the multiple continents. Through the advancement in technology, universities and colleges are now able to create virtual tours that can be accessed across mobile, desktop, and VR headsets. That means if a student is in Beijing, one can check out a university in Texas and feel as if one is really there, just by using a smartphone.

According to Paterson (2018), about 46% of universities and colleges in the United States have deployed virtual reality in some form on the campus, including student recruiting. With the VR technology improving and prices for VR development falling, institutions will find it easier to move ahead. As more K–12 schools are using the virtual devices and that most of the incoming freshman class will have had experience with VR technology, students will expect VR access and to better understand the institution.

THE BENEFITS OF THE VIRTUAL COLLEGE TOUR

More than six million people take a virtual college tour on any given day, which means internet users enjoy the immersive experience a virtual tour provides (Renton, 2019). Virtual college tours make it possible for campuses and universities to market themselves more effectively to an eager student population. Virtual tours can be a benefit for both the student and the college, allowing students to visit campuses without any financial outlay and allowing colleges to market themselves much more effectively to prospective college students.

Virtual college tours make it possible for students to explore college campuses from the comfort of their own homes, thus reducing the number of universities they have to visit in real life. More colleges are expanding their use of virtual reality tours to include different programs of study. In addition, the experiences virtual college tours offer can help students and their families make more informed decisions on the courses and majors they choose. Students on the virtual tour can visualize areas where classes could be in session and experience real dorm rooms and living areas.

Students can be immersed in campus and classroom life, allowing colleges to display their campus culture, facilities, and geographic location to draw students into the local community. Hundreds of schools now offer virtual tours, with options ranging from campus photos that users click through, to elaborate student guides that show off facilities and introduce prospective applicants to peers (Moody, 2019). A virtual campus tour allows the school's administration the ability to display what makes that college different than the rest of the collegiate world.

The more diverse a college campus, the better learning environment for the students, especially when the student population includes students from around the world. Virtual campus tours allow prospective international students from a variety of socioeconomic conditions the ability to see themselves in a college atmosphere. With the help of VR, international students who are interested in knowing more about the college can take a close-up look at the college's learning facilities as well as their living facilities.

Students are able to see campus grounds, accommodation options, classroom and lecture hall sizes, dining halls, and other services from their homes through virtual reality headsets. Not only do the international students benefit from the virtual tour experience, but family, friends, and teachers can also participate if they wish. Virtual tours are certainly essential for international students and parents who might not be able to afford the time or resources needed to visit a campus prior to admittance. Prospective international students can save thousands of dollars by taking advantage of virtual campus tours.

THE SAVANNAH COLLEGE OF ART AND DESIGN (SCAD)

The Savannah College of Art and Design, or SCAD, is a private, nonprofit, accredited institution offering bachelors' and masters' degrees at various locations to prepare talented students for professional careers. SCAD has campus locations in Savannah and Atlanta, Georgia; Hong Kong; Lacoste, France; and online through SCAD eLearning. The diverse student body consists of nearly 12,000 students from across the United States and more than 100 countries.

SCAD was among the first institutions to adopt virtual reality technology. In 2015, SCAD mailed Google Cardboard viewers to 30,000 students who had been accepted to the college but not yet enrolled. Using the Cardboard headset and a smartphone, the students were able to explore the college's campuses in Atlanta and Savannah, Georgia, Lacoste in France and Hong Kong. SCAD saw an enrollment increase of 26% in 2015 through the use of virtual tours, an augmented reality course catalog, and a virtual reality student acceptance letter.

On October 31, 2016, the Savannah College of Art and Design (SCAD) announced the debut of the first virtual reality (VR) musical at that year's Savannah Film Festival. A short film was created entirely by SCAD students and is the first musical performance produced for a virtual reality experience. Created within a Collaborative Learning Center (CLC) course, the project represents 14 of the university's academic programs, including film and television, animation, costume design, dramatic writing, production design, visual effects, themed entertainment design, and motion media design (Falkenberry, 2016).

For filming purposes, SCAD students created a custom camera rig that combined five Sony FS7 cameras and six GoPros to capture every angle on set. With the hopes that virtual reality will help form a stronger community, SCAD plans to create VR videos that showcase the university's top events, galleries and exhibitions. This video content will allow students to see, hear and experience different SCAD locations in a 360-degree format from wherever they are located.

TEXAS A&M UNIVERSITY

In 2017, Texas A&M, the largest university in Texas, gave visitors a realistic way to experience campus life from anywhere in the world. Partnering with concept3D, a 3D mapping and virtual tour software company headquartered in Denver City, Texas, A&M launched what may soon to be a new college recruitment standard. Visitors can toggle between a 360 degree-enabled tour and a virtual reality mode for use with a mobile VR headset. Texas A&M's virtual reality tour allows visitors to orient themselves with an in-window map that shows their location and direction allowing the visitors to use the map to jump to and explore other locations within their view.

Additionally, the new VR system gives Texas A&M staff and students full control of the virtual experience, making it easy to add or update tour content at any time. Visitors can explore the grounds of Texas A&M at any time with a virtual tour or become immersed in one of Texas A&M's iconic campus landmarks. Also available through the virtual tour experience are the many options of residence halls the viewers can visit.

WAYNE STATE UNIVERSITY

Wayne State University (WSU) is a public research university in Detroit, Michigan. Founded in 1868, WSU is Michigan's third-largest university. Wayne State University consists of 13 schools and colleges offering more than 350 programs to nearly 27,000 graduate and undergraduate students. University leaders at WSU were challenged to think critically about how to increase their student appeal and how to differentiate themselves in a crowded education institution market.

Out-of-state student enrollment at WSU was low. Understanding that enrollment growth would likely come from out-of-state students, the provost challenged the Information Technology Department, Marketing Department, and enrollment teams to develop an innovative recruitment strategy targeted to students outside Michigan.

The Information Technology Department, Marketing Department, and enrollment teams recognized that virtual reality could provide the opportunity of bringing the campus to prospective students who were unlikely or unable to make an official on-campus tour. The recruitment team envisioned a virtual reality experience that would enable prospective students at college recruitment fairs to immerse themselves in the campus and the surrounding city of Detroit. The experience would allow users to envision themselves as students at WSU, which should in turn drive interest in the university (Yardey, 2019).

The technology team knew the tremendous cost it would require for the college to hand out VR headsets, which are expensive pieces of equipment. As an alternative, the technology leaders suggested Google Cardboard headsets as a risk-free and more cost-effective means to achieve the virtual reality experience. The university purchased 10,000 college-branded headsets to be handed out at recruitment fairs. They also started filming key spots around campus with 360-degree cameras to build a fully functional VR smartphone app.

Once the application was developed with input from faculty, staff and students, recruitment personnel began handing out the headsets at recruitment events. Prospective students download the WSU application to their smartphone and receive a headset at the university's display stand at college fairs. The prospective students then place their phones in the Google Cardboard device, and together the application and the headset provide a fully immersive VR experience. The prospective students are able to walk through campus, visit classrooms, see Detroit, and even stop by graduation.

When the prospective students leave, WSU hopes they take the headsets and the Wayne State brand home with them to show their family and friends. Recruitment fairs indicated that the VR recruitment effort has been a huge success. The VR goggles produced a buzz around the WSU booth as eager prospective students lined up to visit WSU in virtual reality.

However, the real evidence of success would not become apparent until three weeks after the first out-of-state event, when the AVP for Enrollment received a text from a teacher in Chicago, Illinois. The text contained a picture of students showing off their WSU-branded Cardboard goggles to both teachers and their peers. Students were taking turns using the device and looking around the WSU campus.

REGIS UNIVERSITY

Founded in 1877, Regis University is a private institution in Denver, Colorado. Regis University has a total undergraduate enrollment of 3,961 students and a campus size of 90 acres (How does Regis, 2020). In 2014, the university, in collaboration with independent agency Primacy, created a 360-degree immersive virtual reality tour of their 90-acre campus.

Primacy used 3D printed camera rigs that held up to 10 GoPro Hero4 cameras to capture the footage from around the campus before synchronizing and stitching the footage together. Students were then able to view the recorded footage via an Oculus Rift head-mounted display. Without stepping foot on campus, students across the country can view tours of Regis University's classrooms, sports facilities, dorms, and nearby Red Rocks Amphitheatre as well as a 3D flyover of the campus (Leathers, 2015).

The results included saved time and travel expenses for students and parents as well as a higher return on investment for Regis University. Recruiting events attended by its admissions staff decreased by 14%; however, leads captured from those fewer events increased by 13% percent (Podder, 2017). Regis University uses the virtual reality tour for student recruitment on their university home page as well as at college fairs.

PRINCETON UNIVERSITY

Princeton University is located in Princeton, New Jersey ,and is a private institution that was founded in 1746. Princeton has a total undergraduate enrollment of 5,428. Princeton's setting is suburban, and the campus size is 600 acres (How does Princeton, 2020). In 2016, Princeton University launched an interactive virtual tour that allowed visitors from around the world to explore the campus remotely. The tour is available in four languages, English, Korean, Mandarin, and Spanish. It comprises 23 stops highlighting iconic buildings, academic centers, and student and recreational facilities (Buckley, 2016).

The office of admission spearheaded the virtual tour project and worked with the office of communications and the media company YouVisit to produce the tour. This platform was part of an initiative to reach a broad range of audiences in the digital space, including the recruiting of students who use their mobile phones with access to the Internet. The tour can be viewed on a desktop, tablet, and smartphone. "The virtual tour will be an essential tool in our recruitment of a diverse pool of undergraduates," dean of admission Janet Rapelye said. "Students near and far will be able to experience the beauty of Princeton's main campus in a dynamic and engaging manner" (J. Rapelye, personal communication, July 9, 2020).

Princeton students, several of whom are tour guides on campus, are the virtual guides and greet viewers at each stop inside the virtual tour. Students share a short description, with historical tidbits and fun facts, of campus buildings and locations. Inside the virtual tour, viewers may explore the campus via a walking path, follow the tour from start to finish or jump from stop to stop, and read the location descriptions in the information boxes at the bottom of the screen. Viewers also can experience the tour through virtual reality on their mobile phones with virtual glasses, like Google Cardboard glasses.

Highlights of the tour include:

- The athletics complex, with an imposing interior shot of Jadwin Gymnasium and a 360-degree panorama of the Shea Rowing Center
- East Pyne courtyard and its full interior view of the octagonal Chancellor Green Rotunda
- Engineering Quadrangle, which includes a 360-degree photo of the newly opened Andlinger Center for Energy and the Environment
- Icahn Laboratory and its two-story louvers that move with the sun.

According to Buckley (2016), Rapelye explained "The tour will serve many audiences, including students whose parents might not speak English, international students and students who might find the cost of traveling to campus prohibitive" (J. Rapelye, personal communication, July 9, 2020).

CASTLETON UNIVERSITY

Founded in 1787, Castleton University is a public institution located in Castleton, Vermont. Castleton has an undergraduate enrollment of approximately 1,890. Castleton's setting is rural, and the campus size is 165 acres. In 2018, Castleton University partnered with YouVisit, an interactive content engagement solutions company, to compile a virtual walking tour designed to enhance student recruitment. Making it easier for people both near and far to get a glimpse of the Castleton experience, Castleton University released the final product in spring 2018 (How does Castleton, 2020).

The virtual campus tour allowed prospective students, parents, and other interested parties to tour Castleton in virtual reality. The tour includes 15 different stops, highlighting campus destinations such as the Castleton Fine Arts Center, campus center, Spartan Arena, TV studio, athletic training rooms, dining halls, and dorm rooms. Available in both English and Mandarin, each stop is complete with tour guide audio and features a 360-degree view of each space, with six interactive pieces embedded within the tour.

In 2020, Castleton continues to be a leader in immersive technology. Castleton has transitioned its in-person accepted students' day into an online event on Zoom, allowing the audience to ask questions of current students and other guest speakers (Quigley, 2020).

THE COST OF IMPLEMENTING A VIRTUAL REALITY EXPERIENCE

The virtual reality market is projected to reach $30 billion by the end of 2020 (Burch, 2016). While tech giants and gaming powerhouses may very well dominate the field, startups by the score will no doubt seize their fair share of market revenue. However, in 2020, some colleges are hosting their own virtual platforms. The cost of VR hosting can vary depending on the number of 3D or 2D objects uploaded to the platform and the monthly internet traffic the platform generates.

Most likely a college will create a formal project with an augmented reality (AR) application as the final project outcome. If the college has on-site project managers and VR developers, the college can count their time as indirect cost. However, there are many colleges who will have to outsource the project manager as well as the VR developer. As for the content of a VR application, in some cases, 3D images or videos may be sufficient for virtual tours or e-commerce applications.

Depending on the complexity of a 3D model, development of the 3D model may take between 80 and 400 hours with the hourly rate for the application developer in the range of $80 to $120. The time required to build a VR application can vary in the range of 500 to 2,000 hours. Simple applications with low interactivity and no 3D models will be the fastest to build, while large VR games and simulators with a lot of custom graphics may stretch for about a year to complete. There will be software support that should be considered in the VR application cost (Morozova, 2018).

Regarding the cost for the student, all they need is a VR headset. Using a virtual reality headset such as Oculus Rift, Google Daydream, or Google Cardboard, tour visitors can look around campus, feeling like they're standing right there. Google Cardboard costs less than $20, making it an ideal starter headset for universities who want to encourage prospective students to try out their virtual reality tour. Google's VR View allows one to embed virtual reality content on the website, making it accessible to site visitors from around the world.

THE EMERGENCE OF MIXED REALITY TECHNOLOGY

Mixed reality is a catchall category for all real-and-virtual combined environments generated by computer technology and wearables. Mixed reality (MR) includes two primary types of technology: virtual reality (VR) and augmented reality (AR). Augmented reality superimposes digital information onto the real world. When one puts on a Google Glass or uses a Snapchat filter, AR technology simply alters the experience of the real world, rather than removing it completely. Mixed reality encompasses technology that exists in between VR and AR.

By creating mixed reality tours of the campus, classrooms, athletic facilities and dorms, colleges and universities can emphasize their strengths to prospective students on their websites, at college fairs, high school recruitment visits, and other events geared toward potential students. For example, prospective students can put on a pair of 3D smart glasses with motion-tracking sensors and immediately find themselves in the middle of a college campus, walking along the quad.

Looking to the left, the prospective students might see a trio of modern dorms, while up ahead, they might see the physics building. The students might then virtually walk into the physics building and explore a lab and a lecture hall. If something sparks their interest, they can zero in on it using a hand gesture. If they want more information, they might activate the audio feature, or use hand gestures to open a door to a classroom to see what is inside. Creating immersive AR/VR campus tours is just good business for higher education institutions.

Glasses that allow users to experience virtual and augmented reality have come a long way in the past few years. Today, these fairly lightweight smart glasses have powerful processors, sharp OLED digital displays, motion-tracking sensors and high-resolution cameras, allowing users to experience what they see in 360 degrees.

One college that has been a leader in the use of technology is Orange County Community College, a public community college with two campuses, one in Middletown, New York, and one in Newburgh, New York. In 2019, Orange County Community College implemented North America's first student recruitment augmented reality (AR) campaign, thanks to AR platform, Imagination Park Technologies.

Using Imagination Park's XenoPlay application, students will be able to point their phones toward logos, signs, and other landmarks on campus and reveal exclusive school news, updates, coupons, and more. SUNY Orange is the first college in North America to use Imagination Park's augmented reality to improve campus experiences for students. SUNY Orange's mixed reality projects will help engage and retain students.

Through the Imagination Park XenoHolographic product suite, SUNY Orange faculty, staff, and students will gain access to a content management system, automatic converter of physical objects into AR objects for Android and iOS, virtual AR-based scavenger hunts, and other interactive features. These tools will allow staff and students to develop new VR and AR applications that can be used to maintain their current MR applications and create new immersive applications for different business areas of the college including virtual student recruitment and retention applications.

CONCERNS WITH RECRUITMENT USING MIXED REALITY TECHNOLOGIES

Like any new technology, mixed reality requires staff to invest in the time needed to learn the technology. While prices are starting to fall, some tools for using mixed reality can be expensive to purchase, house, maintain, and keep secure. Colleges may need to be renovated to accommodate the use of mixed reality. Faculty may be reluctant to adopt mixed reality in their testing methodologies.

Research on the effectiveness of mixed reality for learners is only beginning to be developed and industry standards are inconsistent. As with many technologies, mixed reality raises issues about what kind of data generated and who has the right to access that data. Concerns have also been raised that some mixed reality experiences may be so intensely emotional or disturbing that they raise ethical questions about the responsibilities of those who provide the experiences.

VR headsets normally connect to the local WIFI connection. If there are technical issues with internet speed or your VR technology is not working properly, viewers can experience some delays and glitches in the virtual campus tour. This can be eliminated by making sure the institution has a secure and fast Wi-Fi network and updated VR headsets.

There are some critics who see virtual reality campus tours as impersonal. Although the MR technology is advancing, one can appreciate all the work done put in for viewers to experience the VR tour; however, it is not quite the feeling of actually being there. A personal campus visit allows the prospective student to not only get a feel for the atmosphere, but also have a chance to interact with students, faculty, the tour guide, admissions officers, and even professors.

If one is physically on campus, there is a more personalized look into how the college works. There may be an opportunity to sit in on a class, speak to current students about their likes and dislikes of the school, walk around campus to discover class buildings and local shops, and much more. It can sometimes be hard to glean that sort of valuable information from a computer screen or through a VR headset.

A virtual tour may give a great experience of what the campus and the buildings look like; however, if a prospective student relies on just the virtual tour, they might be disappointed when they arrive to register for classes. A virtual tour may not present where the college is located within city or the actual city itself. Some students are drawn to the busy, active vibe of a big city while others would prefer to study in a peaceful countryside setting. Some students would love to see fall colors paint the trees on campus while other students want to get as far away from the possibility of snow as they can.

Virtual tours and photos can only take one so far, and descriptions of colleges may be biased for promotional reasons. The best way to get a genuine sense of the place is to explore it in person. It may be a good idea to do on-site visits to the favorite top three colleges found through viewing the virtual tours. Students get to experience being on multiple campuses before they narrow down their decision. This way, when one eventually does travel to do real on-site college visits, one will have a short list of colleges to visit, saving time and money. Plus, the entire family gets to come along on the preliminary virtual tours.

When using a virtual reality headset, the viewer can engage with the virtual campus in ways that simply are not possible with a traditional mobile or desktop device. Viewers can pan around in any direction and focus their attention as they see fit, just as they could if they were physically standing on campus. While it is possible for viewers to mimic that same freedom by dragging their cursor on a screen, the experience provided by wearing a headset and simply moving their head is far more intuitive and immersive.

CONCLUSION

Finding ways to engage prospective students virtually is important to any long-term marketing strategy. Virtual technology can bring dozens of benefits to almost any field, including higher education. In the era of digital devices, there is an opportunity to enable better recruiting with virtual technology. Virtual reality (VR) seems to be the natural next step for the evolution of higher education recruiting. Many colleges have already taken that next step.

Some colleges have embraced the immersive technologies and have implemented mixed reality in their recruiting processes. When it comes to choosing a college, students have a lot of choices. Finding a way to excite students about college campuses no longer requires expensive, time-consuming site visits. By offering virtual campus tours that emphasize the most important and exciting aspects of campus life, colleges and universities will have their pick of great candidates.

A virtual campus tour is an effective way to draw potential students and their families into a review of the college's special corner of the world, and perhaps to a lifelong relationship between the student and the college. VR has matured and is revolutionizing colleges, offering access to low-income students, experiencing professors who live on other continents, and serving people with disabilities. Colleges are competing for student enrollment. Schools that have taken the plunge and built true virtual tours have reaped the benefits.

According to YouVisit CEO Abi Mandelbaum, universities and colleges his company has worked with have reported an expansion of their reach to prospective students in more than 100 countries, and up to a 30% increase in physical campus visits, leading directly to increased enrollment (The power, 2012). It is time for all education leaders to showcase their colleges and embrace immersive technologies that can enhance their college recruitment.

REFERENCES

Buckley, M. (2016). Virtual tour lets viewers around the globe explore Princeton's campus. Retrieved July 16, 2020, from https://www.princeton.edu/news/2016/03/17/virtual-tour-lets-viewers-around-globe-explore-princetons-campus

Burch, A. (2016, September 29). The top 10 companies working on education in virtual reality and augmented reality. Retrieved July 16, 2020, from http://touchstoneresearch.com/the-top-10-companies-working-on-education-in-virtual

Falkenberry, A. (2016). SCAD students create first virtual reality musical short. Retrieved July 16, 2020, from https://www.scad.edu/about/news-press-and-recognition/2016-10-25-scad-students-create-first-virtual-reality-musical-short

How does Castleton University rank among America's best colleges? (2020). Retrieved July 16, 2020, from https://www.usnews.com/best-colleges/castleton-university-3683

How does Princeton University rank among America's best colleges? (2020). Retrieved July 16, 2020, from https://www.usnews.com/best-colleges/princeton-university-2627

How does Regis University rank among America's best colleges? (2020). Retrieved July 16, 2020, from https://www.usnews.com/best-colleges/regis-university-1363

Leathers, K. (2015, August 19). Regis University creates remote campus tours with Primacy's virtual reality experience. Retrieved July 16, 2020, from https://www.businesswire.com/news/home/20150819005674/en/Regis

Molnar, M. (2019, April 16). Education seen as strong market for VR and AR by industry insiders. Retrieved July 16, 2020, from https://marketbrief.edweek.org/marketplace-k-12/education-seen-strong-market-vr-ar-industry-insiders/

Moody, J. (2019). How to make the most of virtual college tours. Retrieved July 16, 2020, from https://www.usnews.com/education/best-colleges/articles/2019-07-30/how-to-make-the-most-of-virtual-college-tours

Morozova, A. (2018, June 13). How much does a virtual reality app cost in 2018? Retrieved July 16, 2020, from https://jasoren.com/how-much-does-a-virtual-reality-app-cost-in-2018/

Paterson, J. (2018, June 11). Report: 46% of colleges employ virtual reality in courses. Retrieved July 16, 2020, from https://www.educationdive.com/news/report-46-of-colleges-employ-virtual-reality-in-courses/525349/

Podder, A. (2017, June 27). Primacy unlocks philanthropic potential of virtual reality with GivLuv. Retrieved July 16, 2020, from https://mysocialgoodnews.com/primacy-unlocks-philanthropic-potential-virtual-reality-givluv/

The power of virtual tours in student recruitment. (2012, March 23). Retrieved July 16, 2020, from https://monitor.icef.com/2012/09/the-power-of-virtual-tours-in-student-recruitment/

Quigley, A. (2020, April 6). Colleges adjust to virtual recruiting during coronavirus. Retrieved July 16, 2020, from https://vtdigger.org/2020/04/06/colleges-adjust-to-virtual-recruiting-during-coronavirus/

Renton, K. (2019, August 20). Virtual college tours: Boost enrollment & donor engagement. Retrieved July 16, 2020, from https://www.ecityinteractive.com/blog/virtual-college-tours-boost-donor-engagement

Yardey, D. (2019, May 1). How Wayne State University used virtual reality to power student recruitment. Retrieved July 16, 2020, from https://eab.com/insights/blogs/it/how-wayne-state-university-harnessed-virtual-reality-for-student-recruitment/

Chapter 7

Athletics

Laneika K. Musalini

How does the use of virtual reality technology make something feel so real? That feeling is referred to as psychological presence, the science of being there. Virtual reality (VR) is a powerful tool that allows people to explore an environment without actually being in the physical environment. When psychological presence happens, one's motor and perceptual systems interact with the virtual world in a manner similar to how these systems do in the physical world (Bailenson, 2018). VR is evolving and becoming more sophisticated. Soon, almost every household will have some type of VR technology.

When it comes to college athletics, courting top recruits can be extremely expensive. According to the NCAA Financial Report, during the 2019 fiscal year, Georgia became the first school to surpass the $3 million mark in college football recruiting expenses, investing over $3.6 million into football recruiting (Wittry, n.d.). Kentucky eclipsed the $1 million mark spending over $1.07 million on men's basketball recruiting. Recruiting expenses may include transportation, personal aircraft, lodging, and meals for prospective student athletes and their parents, as well as for institutional personnel, telephone call charges, postage, and the list can go on (Wittry, n.d.).

Since the cost of recruiting top-ranked elite college athletic prospects is on the rise, VR could be the answer to cutting recruiting expenses for college athletic departments. In fact, VR technology is on the rise in college athletics. Virtual reality systems are expensive but are seen as an investment for college athletic departments, especially since VR can be multiuse. Additionally, the return on investment is priceless, as some institutions have proven at the tune of national championships. In addition to recruitment, VR technology can be used for athletic training purposes and athletic rehabilitation purposes.

VR IN RECRUITING FOR COLLEGE ATHLETICS

The University of Michigan's athletic department is touted as being the first institution of higher education to use VR technology for recruiting and fan engagement. Michigan's immersive VR experience allows users to see and hear what they would experience if they were actually in the physical environment (Scott, 2015).

Today, VR is taking college campuses by storm. VR is being used as a recruiting tool in a variety of applications—from virtual tours of facilities to virtual point-of-view experiences such as a day in the life of a football player on game day (Schroeder, 2017). To attract talented athletes, the game-day experience can be a recruiter's most powerful tool. All colleges strive to tell a compelling story about their campuses and their offerings that will capture the attention of students, donors, and industry partners. Early adopters are finding that high-tech channels can do that and more, often leading to a stronger bottom line (Bengfort, 2017).

Virtual reality is the newest recruiting essential among college athletic programs. Michigan, Kansas, Georgia Tech, Ole Miss, Syracuse, Mississippi, UNLV, and UCLA are just a few of the athletic programs that have thrown their hats in the virtual recruiting ring (Attwood, 2016). "From a recruiting standpoint, you can show them everything you want to show them, basically have an unofficial recruiting visit from anywhere in the country, as long as they have a smartphone and they can access us on the app," Drew Jennison, UNLV's assistant recruiting coordinator, told the Associated Press (Attwood, 2016, p. 1). This also works for international prospects.

VR IN COLLEGE ATHLETICS

The use of VR in college athletics is a game changer. In addition to using VR for recruiting, college coaches use VR for athletic training and rehabilitation. VR is used as an aid to measure athletic performance as well as to analyze technique and is designed to help improve both performance and technique. The athlete uses VR technology to fine-tune certain aspects of their performance, such as a golfer looking to improve their swing or a track cyclist wanting to go faster in the individual pursuit. Three-dimensional systems can pinpoint aspects of an athlete's performance that require changing things such as the athlete's biomechanics or technique (Virtual Reality Society, 2017).

The use of VR in sports is an immersive experience for athletes. With VR, athletes become immersed in the virtual environment, which allows them to practice under the similar pressures that they experience in a real game. VR also gives athletes an opportunity to practice repetition without exposing their bodies to the physicality experienced during the actual game.

In regards to immersive learning in athletics, VR helps with the mental recall of learning plays and helps athletes absorb plays. According to Omar Ahmad, director of sports partnerships at STRIVR Labs Incorporated, repetition is key. VR puts the player in the moment, explained Ahmad (O. Ahmad, personal communication, 2020). Additionally, immersive learning allows players to practice anytime, anywhere just as if they were on the field, court, slopes, ice, or course. A VR environment allows athletes to accelerate their training regimen, even on the road, giving them the chance to run unlimited repetitions in the most realistic environment possible (Strivr, n.d.).

VR in college athletics is very popular in American football. Many college and university football programs use VR for quarterback training. During the training, quarterbacks are able to identify plays and body movement as well as pressure (O. Ahmad, personal communication, 2020). This type of training increases the quarterback's skills and abilities, such as getting more repetitions in the red zone, identifying plays, and position transition. For football play kickers, VR helps to produce a more accurate kick under pressure.

For athletic rehabilitation, VR helps injured players make good use of their time on the sidelines. After an athlete has been injured while participating in their sport, the last thing they want to experience is another injury. It is difficult to keep athletes engaged throughout the rehabilitation process. In the past, athletes have had to sit out of practices and competitions when they were injured. Many sports are just as much mental as they are physical. With VR, athletes can practice, enhancing their performance and technique while out with an injury. Injured athletes can also review footage, work on timing, and participate in other key aspects of training. VR helps the athlete heal while continuing to enhance their skills and avoiding additional injury.

The University of Alabama introduced VR to athletic rehabilitation in 2017 using the HTC Vive, which has primarily been used for football players but other athletes can take advantage of VR rehabilitation as well (Vicknair, Gsell, & Keith, 2019). The idea of VR rehabilitation was birthed by the Alabama football team's use of VR for athletic training. According to Vicknair et al., incorporating VR into Alabama's practice aligns with their philosophy for attacking rehabilitation: creative, aggressive, and objective. The belief is that this methodology motivates the staff to think outside the box in terms of the tools used and the way rehabs are built and implemented (Vicknair et al., 2019).

As hoped, the rehabilitation team immediately saw immense improvements in athlete motivation and commitment to rehabilitation. Additionally, research has shown that athletes who think about other tasks or places during their rehabilitation can be pushed to attain new goals that surpass the walls of the athletic training room (Vicknair et al., 2019). Besides healing injured athletes, the ultimate goal is for athletes to trust their injured body parts again. Athletes must trust their bodies in order to perform at optimum potential.

COLLEGE SPORTS USING VIRTUAL REALITY

Traditionally, athletic teams depended on playbooks, tapes, and so on to memorize plays. However, VR brings an entirely new perspective to play memorization and places players right into the game. Athletic teams are investing millions in VR sports technology mainly to improve and diversify the training methods of athletes. Sports such as soccer, baseball, basketball, golf, and football are investing in VR technology finding it beneficial for rehabilitation, training, recruitment, and fan engagement (Lynch, 2018).

Baseball

Villanova University uses VR technology in men's baseball. During virtual batting practice, batters see more pitches and hone their pitch recognition abilities (Villanova University Media Room, 2018). While in virtual batting practice, players are providing an opportunity for the Villanova researcher community to design new systems leading to a better understanding of how to further enhance a player's vision training experience. The university has a virtual reality environment that the players use along with the Oculus Rift or the Microsoft HoloLens (Villanova University Media Room, 2018).

The project called PITCHvr brought to life the vision of using VR for baseball training. PITCHvr uses Major League Baseball pitch data available from the PITCHf/x database to recreate the motions of a pitched ball, including the path, velocity, orientation and spin of the baseball from the batters' perspective (Villanova University Media Room, 2018). Training experience is limitless for the baseball players. Using VR technology for baseball training strengthens visual acuity, tracks swing motion to increase accuracy of swings, and allows for multiple repetitions without bodily harm to the pitcher (Gallen, 2019).

Additionally, Dr. Jupina, researcher, and his collaborators plan to add more stadium backgrounds, as well as more realistic animations of pitching avatars to enhance the immersive experience. The team will be capturing batting motion during an actual swing using new imaging technology and incorporating a virtual batted ball's motion for animation, obtaining metrics such as launch angle and velocity of the batted ball, and distance traveled (Villanova University Media Room, 2018).

Basketball

Indiana University is becoming well versed in VR for college athletics. In fact, in 2016 Indiana had at least five sports teams using VR technology, including men's basketball and football (Clavio, 2016). As with other sports, players using VR for basketball can practice shooting the ball and accelerate mental preparedness, repetitions, and so forth. By doing this, players are able to increase the accuracy of their shots in and out of field, play better under pressure, and make better passes. The Mark Cuban Center (the Center for Sports Media and Technology) was built inside the renovated Assembly Hall, home of Indiana's basketball program, giving the Hoosiers the distinction of being the first school in the country to use 3D multi-camera technology and virtual reality (ESPN, 2015).

Thanks to the Center, Indiana students took on the task to develop the VR content themselves. Then senior informatics major and intern at the Center, Catherine Onofrey, spearheaded the development of a VR basketball game, which allows basketball players the opportunity to feel as though they are shooting baskets inside of Simon Skjodt Assembly Hall (Indiana University Athletics, 2017). The game has since been made public for fan enjoyment and participation as well.

Football

Georgia Tech has buzzed into immersive technology by offering a virtual tour of the football athletic facilities. The college recruitment environment has never been more competitive. To gain the cutting edge, Georgia Tech enlisted Foundry 45 to develop a virtual reality application that teleports the viewer to dorm rooms, high-tech classrooms, world-class practice facilities, and right onto Tech's football field (Foundry45.com, 2020).

In fact, viewers get to witness one of the greatest game endings in Georgia Tech football history from the sideline. This captivating immersive recruiting places prospective athletes right into Tech's stadium and gives them an opportunity to experience the overall feel of campus without ever visiting.

Iowa State University is another institution using VR as a recruiting tool. To address the inability to place prospects into the high-energy atmosphere experienced on game day and improve the recruitment process, Iowa's athletic department collaborated with the Virtual Reality Applications Center at Iowa (Kalivarapu et al., 2015). The athletic department and the Virtual Reality Applications Center built an alternative to the game-day experience using the world's highest-resolution six-sided virtual reality environment. Although nothing can replace the actual game day, the immersive recruitment experience gets prospects close.

The training aspect of VR has become a big hit across college football, with teams using it to game plan against opponents; work on specific plays or defensive looks; and give players mental repetitions even when playing at home (Associated Press, 2015). Football coach Mike Norvell of Florida State University implemented the use of VR in football training at Florida State to get a winning edge over opponents. Norvell's belief is that passion and intuition must mesh with analytical thinking to find any advantage possible given that victory so often resides within the margins (Sonnone, 2020).

Norvell strongly believes that VR technology is the key to getting an advantage over opponents. Pairing VR with analytics provides data that is beneficial to performance improvement and overall player development. Norvell knows that VR will allow his players to legitimately take mental repetitions and develop recognition skills away from the gridiron (Sonnone, 2020).

As one can imagine, VR technology in football training is growing rapidly. Syracuse began using VR technology for training in 2015, becoming the first ACC team to do so (Adelson, 2015). Ole Miss, UCLA, and Kansas were also using VR technology in 2015, according to Adelson (2015). EON Sports VR CEO Brendan Reilly believed that the most value was at the quarterback position; however, teams are now using VR technology for the offensive line as well (Adelson, 2015).

Video Gaming and E-Sports

Robert Morris University was the first college to make video gaming a varsity sport. In 2015, Harvard and Stanford joined Robert Morris, in addition to more than 450 other colleges with over 10,000 players comprising The Collegiate StarLeague. This collegiate league even has an annual, major spring tournament where each member of the "Final Four" championship team receives $30,000 in scholarship money (Gregory, 2015).

Since college gaming is not governed by the NCAA and its strict definition of amateurism, e-sports athletes can get cash and other direct benefits from companies (Gregory, 2015). Competitive video gaming is a real thing. According to Earvin "Magic" Johnson, professional video gamers earn seven figures annually. How does he know? He has personally invested in this venture!

COLLEGES USING VIRTUAL REALITY FOR ATHLETICS

Over the last six years or so, more and more colleges have adopted the use of VR technology in college athletics. Collegiate sports are all about competition, development, and family. Most of the top ten college football teams use VR technology for training, rehabilitation, or recruitment. Hundreds of college athletic departments are using VR for a number of things, in addition to those already named, and the list ranges from Division I to junior college.

Some institutions have the capacity to design and produce footage to be used for VR. Other institutions may outsource the service. Iowa State University has the Virtual Reality Applications Center (VRAC). The VRAC is an interdisciplinary research center focused on the intersection of humans and technology, aimed broadly at enhancing the productivity and creativity of people (Virtual Reality Applications Center, 2020).

Stanford University has the Virtual Human Interaction Lab directed by Dr. Jeremy Bailenson, an over 20-year VR industry veteran. The mission of the Virtual Human Interaction Lab is to understand the dynamics and implications of interactions among people in immersive virtual reality simulations, and other forms of human digital representations in media, communication systems, and games.

VIRTUAL REALITY DEVICES

There are a number of companies specializing in VR technology for college athletics. VR is not an augmentation of a previously existing medium, such as adding 3D to movies or color to television. VR is an entirely new medium, with its own unique characteristics and psychological effects, and it will utterly change how people interact with the real world and with other people (Bailenson, 2018).

Recruiting

1. 5 Star Virtual, a VR company, is comprised of 3D virtual tour specialists. 5 Star Virtual provides immersive virtual reality solutions for all types of clients. An impressive project completed by 5 Star Virtual is that of the East Carolina University (ECU) baseball facilities. The virtual tour gives anyone from recruits to fans a behind-the-scenes look at ECU's top-notch facilities without stepping on campus. Fortunately, ECU's project was completed in winter 2019 and launched right before the COVID-19 pandemic in 2020, which forced a global quarantine and caused the cancellation of all spring sports. The virtual tour allowed Coach Godwin and his team to continue recruiting though operations were at a halt.
2. 5 Star Virtual was founded by Mike Mullis, former assistant baseball coach for Pitt Community College and serial entrepreneur. Mullis originally used the VR equipment for his construction and

restoration business, but he quickly saw the practicality of utilizing VR as a recruiting tool. As a former coach, Mullis knew that VR tours for recruiting could save colleges a tremendous amount of money. (5 Star Virtual, 2020)

Training and Rehabilitation

STRIVR Labs Incorporated was founded by Derek Belch and Dr. Jeremy Bailenson. Belch, a former star quarterback at Stanford University, was working as an assistant football coach for the Stanford football team while pursuing a graduate degree in media studies, specifically VR. Bailenson was Belch's graduate adviser. Bailenson, a Stanford professor and founder of the Stanford University Virtual Human Interaction Lab, has spent his career exploring how VR affects the human brain. Bailenson's research has validated immersive technology as an effective medium for learning by driving real-world behavioral change (Strivr, n.d.).

STRIVR's goal is to reinvent and shape the future of how people learn and train. STRIVR has provided VR services for a number of colleges and professional sports teams. STRIVR provides a solution for teams to practice repetitions via VR. The company has worked with players such as Kevin Hogan at Stanford, Deshaun Watson at Clemson, Patrick Mahomes, Carson Palmer, Case Keenum, Ian Mahinmi, skiers of the U.S. Olympic Ski Team with the U.S. Ski & Snowboard, and more (O. Ahmad, personal communication, 2020).

In fact, after using STRIVR's VR technology, Stanford's football team improved their red zone success rate from 50% to 100% in their last 27 trips to the red zone in 2014 (Bailenson, 2018). Hogan improved as a quarterback, as well, after using VR. His pass-completion numbers went from 64% to 76% and the team's total offense improved from 24 points per game to 38 points per game (Bailenson, 2018).

STRIVR's client list currently includes NFL teams, college programs and as of late, NBA, MLB franchises, NHL, WNBA organizations, U.S. Olympic Ski Team, and internationally the German National Soccer Team. In addition to sports, STRIVR also works with industries such as retail, logistics, financial services, manufacturing, and technology for VR development.

Another VR athletic training provider is EON Sports. EON Sports is a division of EON Reality, a global company for Augmented and Virtual reality academic and industry training solutions. EON Sports provided VR for a variety of athletic purposes (EON Reality, 2020). In 2016, EON Sports launched a VR channel with Penn State athletics, called LionVision VR, to enhance fan and alumni engagement. The channel provides fans with behind-the-scenes views of practice, pre- and postgame locker room coaches' talks, and a wide array of iconic Penn State experiences, while immersed inside a customized virtual reality environment (EON Reality, 2016).

In addition to fan engagement, EON Sports provides VR for athletic training. EON's VR sports training platform helps athletes optimize practice and serves as a sophisticated tool for player development, advance scout, and game preparation, allowing players to focus on key performance parameters that help improve their game (FG Team, 2017).

In addition to STRIVR and EON, VAR provides VR services for athletic training. VAR Football is a virtual and augmented reality training suite that allows football players at every position to get extra repetitions to help build comfort and familiarity within a system without hours of physically grueling practice. Developed by Austin James Smith, VAR Football uses two main pieces of technology—HTC Vive for the virtual reality training piece and Microsoft HoloLens for augmented reality training (David, 2017).

VR Devices

Oculus VR, founded by a 21-year-old self-taught engineer, was purchased by Facebook for over $2 billion in 2014 (Bailenson, 2018). Oculus research has a goal to make the consumer virtual, augmented, and mixed reality experiences over the next five to ten years. Oculus is building a team capable of leapfrogging VR forward by bringing together researchers and engineers spanning a wide range of expertise (Oculus, n.d). The Oculus offers several products in its VR line to include Oculus Quest, Oculus Rift S, Oculus Go, Oculus Rift, and Oculus Gear VR. Oculus also powers other devices from competitor product lines.

HTC Vive was released in 2016. Developed in collaboration with Valve, the HTC Vive is a virtual reality headset offering a full-room scale experience. The HTC Vive device uses a gyrosensor, accelerometer, and laser position sensor to track the head's movement as precisely as one-tenth of a degree, in addition the HTC series comes with wireless controllers (D'Ozario & Savov, 2015). Though the Vive has phased out, HTC has enhanced its technology and now offers a Cosmos Series and Pro Series.

The Cosmos Series is the world's first PC-VR system with a suite of modular options for a wide-range of VR applications, and it offers the highest VIVE visual resolution to date (Vive, 2020a). The Professional Series is an easy-to-deploy system that scales with businesses and provides a rich feature set for hard-core gamers. From seated environments to expansive, multiuser deployments, VIVE Pro delivers high-quality visual fidelity, powerful audio, and an immersive experience (Vive, 2020b).

Microsoft offers the HoloLens. The current product, HoloLens 2, offers a comfortable and immersive mixed reality experience with industry-leading solutions that deliver value in minutes. The HoloLens 2 touts reliability, security, and scalability of cloud and AI services from Microsoft (Microsoft, 2020). As if virtual and augmented reality were not enough, the HoloLens 2 is Moore's Law personified.

The HoloLens 2 coupled with software solutions is a hands-free computer that goes where you go, sees what you see, and does what you say, according to Microsoft (2020). Additionally, the device displays a layer of digital information on top of the real world by rendering high-definition holograms that remain where they are put and respond like physical objects when one interacts with them (Microsoft, 2020). This is equivalent to something seen in the movie *Black Panther*.

Google offers Daydream and Cardboard. Though simple and not designed for training or rehabilitation, Google's devices provide an economical option for the beginner, which would work for basic fan engagement or recruitment. Google Daydream is known as simple, high-quality mobile VR that allows one to be immersed in new experiences.

BENEFITS OF VIRTUAL REALITY IN COLLEGE ATHLETICS

Collegiate athletic coaches in all divisions should seriously consider VR technology. For athletic prospect recruiting, VR allows prospects to step into a team's locker room and view what happens during the pregame. Prospects can walk into the weight room or training room and can tour the facilities without ever leaving their home. VR technology for recruiting seems to be working favorably for college athletic departments, especially for coaches who recruit internationally. Additionally, VR technology is the perfect tool for recruiting during COVID-19 because coaches are unable to offer in-person recruitment visits.

VR is far more psychologically powerful than any medium ever invented and is poised to dramatically transform lives. It allows one to instantaneously conjure experiences at the click of a button (Bailenson, 2018). Overall, VR is a way to train players without increasing wear and tear on the players' bodies. Bailenson (2018) further explains that VR is a constant technological system that tracks body movement and updates the sights, sounds, and touch based on those movements; the user feels like they are mentally transported into a different place.

The appeal of virtual reality is one thing; it is clear that the use of this technology gives players an advantage over those whose teams have yet to embrace it (Zorowitz, n.d.). However, one must be aware of the potential side effects associated with VR. STRIVR recommend that athletes only use VR up to 30 minutes at a time (O. Ahmad, personal communication, 2020). Side effects may include anxiety, nausea, and eyestrain all related to “cyber-sickness”; additionally, some people may experience seizures (Smith and Burd, 2019). Yet, an additional benefit is that colleges can now use VR goggles on the sidelines to diagnose concussions (Berger, 2018).

Moreover, sports such as tennis, golf, and soccer could also benefit from the use of VR technology for athletic training. In soccer, defensive players would be able to better prepare for high-performance strikers; strikers could better prepare for defensive tendencies, and even goalkeepers could learn to predict how a striker may approach scoring a goal. In golf and tennis, coaches could build players’ mental preparedness by focusing on how the player can adjust their technique depending on the opponent, course, or court (Christoffersen, 2019).

NCAA

The NCAA rules for Division I college recruiting inhibitsathletic prospects from having communication, recruiting conversations, or verbal offers before June 15 of the prospect’s sophomore year in high school. In fact, prospects are not allowed to have any official or unofficial visits until August 1 of their junior year (NCSA, 2019). Additionally, Division I college recruits are only allowed five official visits at one visit per institution, which can occur during the high school senior year. Further, coaches are permitted three in-person off-campus recruiting contacts (NCSA, n.d.).

Waiting too late to begin unofficial visits can be detrimental for an athletic prospect. The development of VR could help this situation; however, with the NCAA manual being updated annually, the rules may shift to regulate the use of VR in college athletic recruiting (Davis, 2018). Coaches can appreciate the fact that the use of VR technology for recruiting is similar to taking an official visit without coming to campus. As the national governing body of college athletics, the NCAA rules currently require VR applications to have a purpose beyond just recruiting football players (Bengfort, 2017). This is where training and rehabilitation come in, and potentially intercollegiate collaborations.

From a training aspect, supervised virtual reality sessions are included under students’ countable hours, essentially making them the same as traditional film study under the rules, an NCAA spokesperson said (Zorowitz, n.d.). VR technology also helps avoid increased contact hours, which the NCAA continues to crack down on. After all, Division I has a 20/8 Hour Rule. The 20/8 Hour Rule allows student-athletes to participate in countable athletically related activities for four hours per day and 20 hours per week during the academic year and eight hours per week out of season.

FUTURE OF VIRTUAL REALITY IN COLLEGE ATHLETICS

Generation Z is the first generation to grow up entirely in the internet age. Actually, Generation Z is pushing colleges to rethink traditional recruiting. Their expectations shape both form and function, and they expect content targeted to and tailored for them, according to Amir Rasool, managing director of higher education research at Hanover Research (Bengfort, 2017). According to Touchstone Research, college students have a natural predisposition to VR technology, as it is similar to gaming (Burch, 2016).

Moreover, the experience of virtual reality is most interesting and appealing to young people—tweens, teens and millennials or ages 18 to 34 (Burch, 2016). Generations to come will also find the immersive experience of VR technology thrilling and attractive. If Moore's Law proves to be true, VR technology will multiply by leaps and bounds over the next 10 years.

Thinking forward, VR will be used for fan engagement. There is no day like game day. By allowing fans to become more immersed into the game, fans can enjoy the sport of their choice up close and personal with 3D images rather than 2D. When fans cannot make it to the actual game or match, they will be able to view the game using VR. Fans will have the ability to take behind-the-scenes tours of the training facilities or a new stadium before it is built, watch practices, rewatch plays from any angle, and view the team's bus and plane camera (Attwood, 2016). This not only enhances fan engagement, but it also allows alumni, parents, and donors to be engaged.

Game day officiating will also be a thing in the future. Video Assistant Referees (VAR) are already being used by FIFA. It is a form of assistance that reduces mistakes, and its decisions must be accepted, even if it takes a bit of time (FIFA, 2020).The purpose of Virtual Assistant Referees is to reduce the number of errors in play calling, which drives more fairness (FIFA, 2020).

CONCLUSION

At least one tech-focused investor group is predicting VR to be a mainstream technology worth an estimated $60 billion over the next decade (Bailenson, 2018). In fact, virtual reality has introduced an entirely new way of being and of doing. Virtual reality offers immense diversity of technology and has potential to expand usability even further.

For college athletic departments, VR has proven to be an innovative solution to recruiting, training, rehabilitation, and fan engagement. Virtual reality is extremely useful in sports that require great bodily contact. Preserving athletes' health and safety is paramount for the athlete and the organization. By using VR, elite athletes avoid additional wear and tear on the body.

Due to its uniqueness and vast applications, VR is now being used by professional athletic associations, such as the NFL, NBA, MLB, PGA, and MLS, and is trending globally. The future of VR in college athletics is bright. More so, the future of technology companies such as STRIVR, 5 Star Virtual, Oculus, Sony, LiveLike, and others is also bright as the companies continuously strive to improve VR.

CHAPTER SUMMARY

- When it comes to college athletics, courting top recruits can be extremely expensive.
- Since the cost of recruiting top-ranked, elite college athletic prospects is on the rise, VR could be the answer to cutting recruiting expenses for college athletic departments.
- The University of Michigan's athletic department is touted with being the first institution of higher education to use VR technology for recruiting and fan engagement.
- Michigan, Kansas, Georgia Tech, Ole Miss, Syracuse, Mississippi, UNLV, and UCLA are just a few of the athletic programs that have thrown their hats in the virtual recruiting ring (Attwood, 2016).
- In addition to using VR for recruiting, college coaches use VR for athletic training and rehabilitation.
- With VR, athletes become immersed in the virtual environment, which allows them to practice under the similar pressures that they experience in a real game.
- Many college and university football programs use VR for quarterback training.
- For athletic rehabilitation, VR helps injured players make good use of their time on the sidelines.
- Sports such as soccer, baseball, basketball, golf, and football, to name a few, are investing in VR technology finding it beneficial for rehabilitation, training, recruitment, and fan engagement.
- The project called PITCHvr brought to life the vision of using VR for baseball training.
- Georgia Tech has buzzed into immersive technology offering a virtual tour of the football athletic facilities.
- The training aspect of VR has become a big hit across college football, with teams using it to game plan against opponents, work on specific

plays or defensive looks, and give players mental repetitions even when playing at home (Associated Press, 2015).

- Robert Morris University was the first college to make video gaming a varsity sport.
- Some institutions have the capacity to design and produce footage to be used for VR. Other institutions may outsource the service.
- There are a number of companies specializing in VR technology for college athletics.
- STRIVR Labs Incorporated was founded by Derek Belch and Dr. Jeremy Bailenson.
- STRIVR has provided VR services for a number of colleges and professional sports teams.
- EON Sports provided VR for a variety of athletic purposes.
- VAR Football is a virtual and augmented reality training suite that allows football players at every position to get extra repetitions to help build comfort and familiarity within a system without hours of physically grueling practice (David, 2017).
- Oculus research has a goal to make the consumer virtual, augmented, and mixed reality experiences over the next five to ten years.
- The Cosmos Series is the world's first PC-VR system with a suite of modular options for a wide-range of VR applications, and it offers the highest VIVE visual resolution to date.
- Microsoft offers the HoloLens. The current product, HoloLens 2, offers a comfortable and immersive mixed reality experience with industry-leading solutions that deliver value in minutes (Microsoft, 2020).
- The device displays a layer of digital information on top of the real world by rendering high-definition holograms that remain where they are put and respond like physical objects when one interacts with them (Microsoft, 2020).
- Google offers Daydream and Cardboard.
- VR technology for recruiting seems to be working favorably for college athletic departments, especially for coaches who recruit internationally.
- Overall, VR is a way to train players without increasing wear and tear on the players' bodies.
- However, one must be aware of the potential side effects associated with VR. STRIVR recommends that athletes only use VR up to 30 minutes at a time.
- Sports such as tennis, golf, and soccer could also benefit from the use of VR technology for athletic training.
- Generation Z is the first generation to grow up entirely in the internet age and is pushing colleges to rethink traditional recruiting.
- VR technology will multiply leaps and bounds over the next 10 years.

- The purpose of Virtual Assistant Referees is to reduce the number of errors in play calling, which drives more fairness (FIFA, 2020).
- For college athletic departments, VR has proven to be an innovative solution to recruiting, training, rehabilitation, and fan engagement.

REFERENCES

5 Star Virtual (2020). Capture the attention of your audience: Showcase your virtual tours on the web. Retrieved from http://www.5starvirtual.com/

Adelson, A. (2015 May 29). Syracuse turns to virtual reality for training. Retrieved from https://www.espn.com/blog/acc/post/_/id/82369/syracuse-turns-to-virtual-reality-for-training

Associated Press (2015 November 21). New uses of virtual reality adds another dimension to coaches' recruiting tactics. *New York Times*. Retrieved from https://www.nytimes.com/2015/11/22/sports/ncaafootball/new-use-of-virtual-reality-adds-another-dimension-to-coaches-recruiting-tactics.html

Attwood, E. (2016). Virtual reality's role in the future of sports. *Athletic Business*. Retrieved from https://www.athleticbusiness.com/apps-software/virtual-reality-s-role-in-the-future-of-sports.html

Bailenson, J. (2018). *Experience on demand: What virtual reality is, how it works, and what it can do*. New York: W.W. Norton & Company, Inc.

Bengfort, J. (2017). College recruiting goes virtual with immersive technology. *EdTech Magazine*. Retrieved from https://edtechmagazine.com/higher/article/2017/10/college-recruiting-goes-virtual-immersive-technology

Berger, M. (2018). How virtual reality and sideline brain scans could help diagnose concussions. *Smithsonian Magazine*. Retrieved from https://www.smithsonianmag.com/science-nature/virtual-reality-sideline-brain-scans-help-diagnose-concussions-180970249/

Burch, A. (2016). VR and consumer sentiment. Retrieved from http://touchstoneresearch.com/vr-and-consumer-sentiment/

Christoffersen, B. (2019). *Using virtual reality to train athletes* (thesis, Concordia University, St. Paul). Retrieved from https://digitalcommons.csp.edu/sport-management_masters/26

Clavio, G. (2016). Athletes look for an edge in a new place: Virtual reality. Retrieved from https://theconversation.com/athletes-look-for-an-edge-in-a-new-place-virtual-reality-56776

David, S. (2017). VAR football ushers in new age of training with holograms in college football. Retrieved from https://www.sporttechie.com/var-football-uses-technology-to-usher-in-new-age-of-football-training

Davis, M. (2018). Take a walk in my headset: A look into the use of virtual reality in Division I football recruiting efforts. Retrieved from https://jolt.richmond.edu/2018/11/30/take-a-walk-in-my-headset-a-look-into-the-use-of-virtual-reality-in-division-i-football-recruiting-efforts/

D'Ozario, D., & Savov, V. (2015). Valve's VR headset is called the Vive and it's made by HTC. Retrieved from https://www.theverge.com/2015/3/1/8127445/htc-vive-valve-vr-headset

EON Reality (2020). About us. Retrieved from https://eonreality.com/company/

EON Reality (2016). EON sports VR launches virtual reality channel with Penn State athletics Nittany Lions fans. Retrieved from https://eonreality.com/company/

ESPN (2015). Mark Cuban donates $5 million to Indiana University. Retrieved from https://www.espn.com/mens-college-basketball/story/_/id/13020731/mark-cuban-gift-provide-indiana-hoosiers-new-video-center

FG Team (2017). EON sports VR offers immersive virtual reality training for football and baseball. Retrieved from https://www.fitness-gaming.com/news/fitness-and-sports/eon-sports-vr-offers-immersive-virtual-reality-training-for-football-and-baseball.html

FIFA (2020). Football technology: VAR at the 2018 FIFA World Cup. Retrieved from https://football-technology.fifa.com/en/innovations/var-at-the-world-cup/

Foundry45.com (2020). Virtual reality experience case study: GT football athletic recruiting. Retrieved from https://foundry45.com/vr-case-studies/georgia-tech-virftual-reality-football-recruiting-case-study/

Gallen, P. (2019 March 29). Villanova University using virtual reality to help baseball team. CBS Philly 3. Retrieved from https://youtu.be/-sn8rIFpgJg

Gregory, S. (2015). The varsity sport of the virtual world. *TIME*. Retrieved from https://time.com/3759634/virtual-world-varsity-sport/

Indiana University Athletics (2017). Cuban Center student develops virtual reality basketball game. Retrieved from https://iuhoosiers.com/news/2017/3/29/general-cuban-center-student-develops-virtual-reality-basketball-game.aspx

Kalivarapu, V., MacAllister, A., Hoover, M., Sridhar, S., Schlueter, J., Civitate, A., Thompkins, P., Smith, J., Hoyle, J., Oliver, J., Winer, E., & Chernoff, G. (2015). Game-day football visualization experience on dissimilar virtual reality platforms. In *Mechanical engineering conference presentations, papers, and proceedings*. Retrieved from https://lib.dr.iastate.edu/me_conf/173/?utm_source=lib.dr.iastate.edu%2Fme_conf%2F173&utm_medium=PDF&utm_campaign=PDFCoverPages

Lynch, M. (2018). How college athletics programs are innovating with VR. Retrieved from https://www.thetechedvocate.org/college-athletics-programs-innovativating-vr/

Microsoft (2020). HoloLens 2. Retrieved from https://www.microsoft.com/en-us/hololens/hardware

NCSA Sports (n.d.). NCAA football recruiting rule and calendar: When can college football coaches contact you. Retrieved from https://www.ncsasports.org/football/recruiting-rules-calendar

NCSA Sports (2019). New NCAA recruiting rules on early recruiting. Retrieved from https://www.ncsasports.org/blog/2019/04/26/ncaa-di-recruiting-rules-early-recruiting/

Oculus (n.d.). Research: Join us in creating the future. Retrieved from https://www.oculus.com/research/

Schroeder, S. (2017). What virtual reality can do for college recruiting. *Athletic Business*. Retrieved from https://www.athleticbusiness.com/apps-software/what-virtual-reality-can-do-for-college-recruiting.html

Scott, J. (2015). Michigan first school to use virtual reality in recruiting, fan engagement. *Athletic Business*. Retrieved from https://www.athleticbusiness.com/college/virtual-reality-coming-to-campus-at-michigan.html

Smith, S., & Burd, E. (2019). Response activation and inhibition after exposure to virtual reality. *Elsevier*, 3–4. Retrieved from https://doi.org/10.1016/j.array.2019.100010

Sonnone, B. (2020). Novell values technology and analytics to find winning edge. Retrieved from https://247sports.com/college/florida-state/Article/FSU-football-coach-Mike-Norvell-analytics-technology--144218030/

Stanford University (n.d.). Virtual human interaction lab mission. Retrieved from https://vhil.stanford.edu/mission/

STRIVR (n.d.). Take your team's training to the next level. Retrieved from https://www.strivr.com/use-cases/sports/

Vicknair, R., Gsell, J., & Keith, C. (2019). Seeing things differently: University of Alabama using VR in athlete rehab. Retrieved from https://training-conditioning.com/article/see-things-differently-d66/

Villanova University Media Room (2018). Villanova faculty making batting practice a virtual reality. Retrieved from https://www1.villanova.edu/villanova/media/pressreleases/2018/0522.html

Virtual Reality Applications Center (2020). About the VRAC. Retrieved from http://www.vrac.iastate.edu/

Virtual Reality Society (2017). Virtual reality in sport. Retrieved from https://www.vrs.org.uk/virtual-reality-applications/sport.html

Vive (2020a). Vive: Cosmos series. Retrieved from https://www.vive.com/us/product/#cosmos%20series

Vive (2020b). Vive: Professional series. Retrieved from https://www.vive.com/us/product/#pro%20series

Wittry, A. (n.d.). An analysis of college football recruiting costs. Retrieved from https://www.athleticdirectoru.com/articles/an-analysis-of-football-recruiting-costs/

Zorowitz, J. (n.d.). It just got real. *NBC Sports*. Retrieved from https://sportsworld.nbcsports.com/virtual-reality-sports-arkansas-kentucky/

Chapter 8

Virtual Reality in General

Lavar N. Marlowe

The world of technology is changing rapidly. Are educators stuck in an antiquated view of technology or are they striving to make the necessary changes to keep up with the world and all that it has to offer? Are they teaching students in old ways of thinking or are they looking ahead to make sure the world does not leave American students in community colleges or universities behind?

VIRTUAL REALITY AND ITS BEGINNING

Over the years, virtual reality has changed significantly. In 1898, stereoscopic photos and viewers were introduced. Charles Wheatstone "demonstrated that the brain processes the different two-dimensional images from each eye into a single object of three dimensions. This new thought led to the View-Master Stereoscope" (Virtual Reality Society, n.d., para. 8). In 1929, the "Link Trainer," the first flight simulator, was created. In the later 1930s, *Science Fiction Story* predicted virtual reality.

In the 1930s, Stanley G. Weinbaum predicted that a set of goggles could create a new experience. He was truly a visionary. Morton Heilig's "Sensorama" was developed and patented in 1962. It was "an Arcade-style theatre cabinet that would stimulate all the senses, not just sight and sound" (Virtual Reality Society, n.d., para. 11). This was a precursor of arcade games developed in the late 1980s and 1990s. During this time Morton Heilig invented the first virtual reality head-mounted display, a forerunner of today's Oculus or Google Cardboard devices.

In 1961, a headset was created that was the first motion-tracking-head mounted display. "Head movements would move a remote camera, allowing the user to naturally look around the environment" (Virtual Reality Society, n.d., para. 13). In 1965, The "Ultimate Display," by Ivan Sutherland, came on the scene that "could stimulate reality to the point where one could not tell the difference from actual reality" (Virtual Reality Society, n.d., para. 14). "In 1968 Ivan Sutherland and his student Bob Sproull created the first VR head mounted display (Sword of Damocles) that was connected to a computer and not a camera" (Virtual Reality Society, n.d., para. 17).

This is significant because technological advancements had moved to a computer, which would substantially change VR. "In 1969 Myron Kruegere, a virtual reality computer artist, developed a series of experiences which he termed 'artificial reality.' He developed computer-generated environments that responded to the people in it" (Virtual Reality Society, n.d., para. 18).

These developments through recent history comprise the basic foundation of virtual reality today. Big business recognized the importance of virtual reality. In 1972, "General Electric produced a 'computerized' flight simulator that sported three screens arranged in a 180-degree configuration" (Virtual Reality Society, n.d., para. 20). In the late 1970s, education played a part in moving virtual reality to its next phase. In 1977, "MIT created the Aspec Movie Map. This system allowed people to wander through a virtual experience of Aspen, Colorado" (Virtual Reality Society, n.d., para. 22).

After years of flight simulators being produced for pilot education, in 1979, "The VITAL helmet was probably the first proper example of a VR HMD outside of the lab. Using a head tracker, pilots could look at primitive computer-generated imagery" (Virtual Reality Society, n.d., para. 23). In 1982, "Finger-tracking gloves for VR called 'Sayre' gloves were invented by Daniel Sandin and Thomas DeFanti." "The gloves were wired to a computer system and used optical sensors to detect finger movement" (Virtual Reality Society, n.d., para. 24).

In 1985, virtual programming language began to become relevant. Here virtual programming language was coined by computer scientists Jaron Lanier and Thomas Zimmerman. "This is the first ever VR company to sell HMDs and gloves. The term 'data glove,' comes from their DataGlove product" (Virtual Reality Society, n.d., para. 25).

In 1987, "Jaron Lanier, founder of the visual programming lab (VPL), popularized the term 'virtual reality'" (Virtual Reality Society, n.d., para. 27). This is where research, data, and science collide. The federal government saw a need, and NASA engineers became a part of the movement in 1989 with a virtual reality simulator that was used to train astronauts. It also led to the power glove being used by Nintendo, which catapulted the gaming world into a whirlwind of virtual reality business.

The next wave was virtual reality group "Arcade Machines" in 1991, along with a coupling of science with Medina's virtual reality Mars Rover. SEGA Corporation created a virtual reality headset for a gaming console produced in 1993. Fast-forward to 2007 with Google bringing Google Street View (Virtual Reality Society, n.d.). "Google enhanced its Maps service with street-level 360-degree images, captured by special cars fitted with custom camera equipment" (Virtual Reality Society, n.d., para. 35).

In 2012, Palmer Luckey developed the Oculus VR. In 2016–2017, "everyone unleashes VR products that are ready for primetime" (Virtual Reality Society, n.d., para. 39). Once something hits mainstream arenas for normal buying and household consumption, the world changes into a receptive mode for the device. As an example, once smartphones became normal, land-line phones became almost obsolete. The same happened with cars with cruise control in the 1950s. Once it became mainstream, no one wanted a car without it.

Virtual reality has been a technology topic for decades. Educators are on the cusp of the next big wave. This is a technology that the higher education systems can use to teach students and create programs. Virtual reality can be a major player in education. In the near future higher education will wonder how it can survive without virtual reality education. Higher education can be one of the leaders in educating its administrators on the importance of being prepared to teach with virtual reality. All should be aware and prepare for this wave of technology.

If student services divisions could use virtual reality, it could literally show students around campus instead of handing them a map and telling them to find it on their own. VR could aid student orientation programs and make them more hands-on. Deliberately immersing students into the campus could increase the retention rates between orientation and starting classes. If virtual reality was used in vocational training, students could actually loosen or tighten the bolts used in an automotive class.

If virtual reality was used in science, it could assist students with the anatomy of the body, provide a 360-degree view of a heart, or demonstrate what goes on in a virtual hospital. If students were able to immerse themselves in virtual reality, they could create photography or videography applications to create films or photos to create stories. It could create access to all programs to and build world-class higher education institutions of learning.

VIRTUAL REALITY CHANGES PERCEPTION

Virtual reality creates something called presence, the state of existing, or being present in a place or thing. Virtual reality allows students to believe that they are present in the situation being provided through virtual reality. It creates a physiological result by activating the brain to think it is in that virtual reality moment. This is important because creating these virtual scenarios can help higher education educate on subjects that a photo and a book could never create. There is heightened memory accuracy. After using a virtual reality tool, one is more likely to remember more than viewing a computer screen.

VIRTUAL REALITY IN THE LIBRARY SCIENCE, TRAINING COURSES, AND EDUCATION

East Carolina University in North Carolina is a leader in the area of virtual reality. In library science, training courses, and education the university is using virtual reality in information literacy instruction to reduce library anxiety among students. Virtual reality will challenge education to dramatically change how data is collected, how students learn, and how faculty teach.

Higher education must improve how it shares training, education, and learning. For example, having students using virtual reality to view slavery, to view the pain of being whipped, and to be shackled is a feeling that could be experienced. This will create empathy as to why it was wrong. Further, it could create learning for generations to come. Reading a book is fine, but being immersed in its content through virtual reality is much more powerful.

Advantages in using virtual reality are highlighted by East Carolina professor Veronica S. Pantelidis. In *Themes in Science and Technology Education*, Pantelidis (2009) explored virtual reality. She gave the following reasons to use virtual reality in education:

1. In some instances, VR can more accurately illustrate some features, processes, and so forth than by other means, allowing extreme close-up examination of an object, observation from a great distance, and observation and examination of areas and events unavailable by other means.
2. Virtual reality motivates students. It requires interaction and encourages active participation rather than passivity.
3. Virtual reality allows the learner to proceed through an experience during a broad time period not fixed by a regular class schedule, at their own pace. It allows the disabled to participate in an experiment or learning environment when they cannot do so otherwise. It transcends language barriers. (Pantelidis, 2009, p. 62)

Some students, when approaching the workforce development area of the college, already know exactly what they want to study. Many of these students are nontraditional, older, and decided to come back for another career. These students may have problems with computers and technology. Virtual reality could ease and comfort these students. Simple and easy, it can aid with instruction, and "grab and hold the attention of the students. Virtual Reality can more accurately illustrate some features, processes, and so forth than by other means" (Pantelidis, 2009, p. 63).

Pantelidis says, "Looking at the model of an object from the inside or the top or bottom shows areas never seen before" (Pantelidis, 2009, p. 63). Virtual reality creates interaction with the source and causes the student to better acclimate with the learning function. Virtual reality "can change the way a learner interacts with the subject matter" (Pantelidis, 2009, p. 63). Training courses and education as a whole would benefit from virtual reality integrated in its curriculum.

PAUL G. ALLEN SCHOOL OF COMPUTER SCIENCE AND ENGINEERING

Another leader in this area is the Paul G. Allen College of Science and Engineering at the University of Washington. A Virtual Reality System course turned students into makers, especially during the COVID-19 pandemic of 2020. "The hands-on course introduced students to the cross-disciplinary skills needed to create their own VR systems" (Paul G. Allen School of Computer Science & Engineering, n.d., para. 2). The graduate students who are already immersed in creating programs were eager and excited about this program.

VITAL

San Diego State University (n.d.) is looking to change academia into a virtual reality learning world. If the university is successful, it can drive an increase in revenue, create a more diverse student body, support faculty members, and gain recognition. In 2017, San Diego State University launched their first virtual reality program called Virtual Immersive Teaching and Learning (VITal).

VITal is set up in various academic disciplines like art, biology, engineering, math, music, and so on, providing students the opportunity to expand in their career area of interest in much greater depth. Imagine traveling to the Louvre Museum in Paris to learn about the Mona Lisa or having the opportunity to conduct open-heart surgery. This is the gateway for students to be able to explore their career paths in greater depth. This opportunity is granting students the ability to experience the possibilities for themselves.

VITal is taking the extra steps for students to equip themselves for their future career endeavors. It is exploring different virtual reality headsets like Oculus Quest, HTC Cosmos, the Windows Mixed Reality headset, and the Standalone viewer in certain academic departments. VITal was first introduced to the faculty before launching to the students. To ensure the success of any program, the university needed the facility to be on board.

Once the technology was learned, it was fairly easy to incorporate in the classroom. Normally, in classroom settings every student was required to bring a variety of items, such as pencils, paper, notebooks, laptops, laboratory gloves, goggles, etc. However, virtual reality only requires the students. The professor can set up the task for what students are required to perform or learn for that particular day. For instance, medical students may start off by learning CPR. During the virtual reality, students may be asked about what tools to use during CPR and to demonstrate how to use the tools.

Not only can the students learn real-life situations, but they can repeat the lesson as often as needed. Students learn more from hands-on learning and repetition as it helps to stimulate the brain to retain the information that was learned. It is the same scenario as driving to a familiar destination, like work. The mind is already trained on how to get to work, because of repetition, which is why individuals are able to drive to work without any issues.

VITal uses apps to attract student's attention to the new virtual reality platform. Apps are a great way to introduce virtual reality to students, as a majority of students have access to mobile phones. It is also a powerful method to connect outside of the university area to other communities, cities, and states. This tactic could potentially be a selling point for the university to advertise their new platform once it has gained success at the university.

Although virtual reality is not real, the simulation is what makes the situation appear real. The realness of the simulations will force students to think, act, and respond quickly in order to resolve the issue in a timely and appropriate fashion. VITal will help train students to stay focused, remain attentive, and to develop new skill sets like communication and listening.

PERCEPTIONS OF VIRTUAL REALITY

Virtual reality teaches students how to visit ancient Rome, work with the human anatomy, or learn law enforcement techniques. Modern virtual reality concepts feel real. The educational world has not even scratched the surface on what virtual reality can do. Virtual reality is set to assist teachers, educators, and leaders into a realm where online education has already skyrocketed to the forefront. Older generations, older educators, or even young educators who do not want to learn to work with a new technology may be apprehensive to use VR technology in higher education.

Higher education can lead companies by teaching them to rely on virtual reality to learn about a product, construct a building, or envision future merchandise before it is designed or manufactured. This would lead to better data-informed decisions. Higher education can help leaders take a glimpse into the future with sports that are more competitive. Virtual reality can help higher educational medical programs guide students in the study of dementia or stress/anxiety.

SIMT IN THE WORLD OF VIRTUAL REALITY

Community College and University Administrators should take a note of Florence Darlington Technical College (FDTC) in Florence, South Carolina. These leaders have proven to be a leader in workforce development and continuing education. This college has the capacity to support its community and students in a manner that is beyond compare.

The Southeastern Institute of Manufacturing Technology (SiMT) has discovered a way to stay ahead of the competition by educating a manufacturing and technology driven workforce. One of SiMT's greatest impacts on business and industry is with virtual reality. SiMT's virtual reality center is phenomenal. "The SiMT 3D Interactive Production Studio is dedicated to creating dynamic, interactive virtual experiences that enhance occupational and safety training, improve product marketing, and provide effective visualizations of construction and production projects for numerous industries" (SiMT, n.d., para. 1).

FDTC attracts business, government agencies, and other higher education programs to use their space with virtual reality platforms like Oculus Rift/Touch, HTE, and Zspace. Companies that use Virtual Reality have discovered the benefits that this economical technology offers in the areas of training, planning and marketing. The SiMT team has developed plant tours, virtual welding, and Forklift operations, which provide safe and highly effective learning environments for employees across industries. "We work with Fortune 500 companies to create virtual planning tools that allow for optimized productions flows and visualization for future construction projects" (SiMT, n.d., para. 2).

GENERAL BARRIERS AND SECOND LIFE VIRTUAL WORLDS

There are barriers to virtual reality. Steven Warburton (2009) wrote an article entitled, "Second Life in Higher Education: Assessing the Potential for and the Barriers to Deploying Virtual Worlds in Learning and Teaching," to examine these barriers. When thinking of Virtual Reality in Education, the term *virtual worlds* is important to note "a virtual world provides an experience set within a technological environment that gives the user a strong sense of being there" (Warburton, 2009, p. 415).

Warburton brings the element of technical issues. Budgets are always a concern, obtaining funding to make sure proper equipment is available to students could be a hardship in the beginning. Most higher education systems are already taxed, and it is hard to purchase the best equipment. Virtual worlds are important to view through the lens of the consumer, which would be best for those in higher education.

Warburton says:

> The multi-user virtual environments (MUVEs) of today share common features that reflect their roots in the gaming worlds of multi-user dungeons and massively multiplayer online games (MMOs), made more popular in recent times through titles such as NeverWinter Nights and World of Warcraft, both based on the Dungeons and Dragons genre of role-playing game. (Warburton, 2009, p. 415)

Second Life is a virtual world to use in higher education. If used appropriately, the technical issues that underlie this particular way of usage for virtual reality can be problematic. The Second Life Virtual World could be used to create an experience for users in higher education that could be simply extraordinary. "The visual experience is rendered in real time, unlike other gaming engines where objects such as monsters and weapons are stored locally" (Warburton, 2009, p. 418).

The world has changed, and higher education should prepare to use virtual reality programs. Lastly, "Access is also a way to deliver new things in close to real time" (Kelly, 2017, p. 114). Most employers cannot meet demand in a good economy and need learned workers immediately. Virtual reality can create on-demand real-time educational resources for the workforce.

VIRTUAL REALITY

Today with the advent of the latest technology, higher education has the opportunity to take the power from something as simple as Google Cardboard to change what happens in the educational world. At first it may feel strange to use something like headsets to lead one's institution to better use of time and lesson explanations. There is surprising joy in the usage of various types of headsets for all ages.

COPYRIGHT RESTRICTIONS

An area to be aware of with using virtual reality in higher education is copyright laws to protect the institution and the one who will design the course. Alexis Dunne (2019) has done a phenomenal job in providing information on this topic. If and when someone decides to create a virtual reality program at an institution of higher education, one should review the specific laws needed to make sure no laws are being infringed upon.

Dunne (2019) contends:

> The source of Congress's ability to grant copyrights to authors of copyrightable works through the Copyright Act is found in Article 1, Section 8, Clause 8 of the United States Constitution. It grants Congress the power "[t]o promote the progress of science and useful arts, by securing for limited times to authors and inventors the exclusive right to their respective writings and discoveries." (p. 339)

Making sure every institution of higher education is protected is an issue that should be examined and understood.

CONCLUSION

Technology has provided a useful tool to aid students and prepare them in their learning for all programs. One must be a visionary, staying up to date with current trends and data pertaining to the community to which it serves. One must acknowledge risks of not adhering to change, and the potential for problems for higher education if it is not organized with the future in mind. Effective leaders within higher education must continuously learn and be prepared to lead in circumstances beyond the scope of their control.

CHAPTER SUMMARY

- Over the years, virtual reality has changed significantly.
- These movements through history comprise the basic foundation of virtual reality which led to what it is today.
- The federal government saw a need and NASA Engineers became a part of the movement in 1989 with a virtual reality simulator was used to train astronauts.
- Virtual reality can be a major player in education. In the near future higher education will wonder how it can survive without virtual reality education.
- Virtual reality creates something called presence, the state of existing, or being present in a place or thing. Virtual reality allows students to believe that they are present in the situation being provided through virtual reality.
- Virtual reality will challenge education to dramatically change how one interacts with data collected, how one interacts with each other, and how one teaches.
- Training courses and education as a whole would benefit from virtual reality integrated in its curriculum.
- In 2017, San Diego State University launched their first virtual reality program called Virtual Immersive Teaching and Learning (VITal).
- Students learn more from hands-on learning and multiple repetition as it helps to stimulate the brain to retain the information that was learned.
- VITal will help train students to stay focused, remain attentive, and to develop new skill sets like communication and listening.
- Higher education can lead companies by teaching them to rely on virtual reality to learn about a product, construct a building, or envision future merchandise before it is designed or manufactured.
- The SiMT team has developed plant tours, virtual welding, and forklift operations that provide safe and highly effective learning environments for employees across industries.
- The Second Life Virtual World could be used to create an experience for users in higher education that could be simply extraordinary.
- The world has changed, and higher education should prepare to use virtual reality programs.
- One must be a visionary, staying up to date with current trends and data pertaining to the community to which it serves.

REFERENCES

Dunne, A. (2019). Copyrighting experiences: How copyright law applies to virtual reality programs. *The Journal of Business, Entrepreneurship & the Law, 12*(2). Retrieved from https://digitalcommons.pepperdine.edu/cgi/viewcontent.cgi?article=1198&context=jbel

Kelly, K. (2017). *The inevitable: Understanding the 12 technological forces that will shape our future*. New York: Penguin Books.

Krokos, E., Plaisant, C., & Varshney, A. (2016). Virtual memory palaces: Immersion aids recall. *Virtual Reality, 23*(1–15). doi:10.1007/s10055-018-0346-3

Pantelidis, V. S. (2009). Reasons to use virtual reality in education and training courses and a model to determine when to use virtual reality. *Themes in Science and Technology Education, 2*(1–2), 59–70. Retrieved from https://files.eric.ed.gov/fulltext/EJ1131313.pdf

Paul G. Allen School of Computer Science & Engineering (n.d.). New virtual reality systems course turns students into makers. Retrieved from https://news.cs.washington.edu/2020/03/31/new-virtual-reality-systems-course-turns-students-into-makers/

San Diego State University (n.d.). Virtual immersive teaching and learning (VITaL). Retrieved from https://its.sdsu.edu/vital/

Southeastern Institute of Manufacturing Technology (n.d.). Interactive production studio. Retrieved from http://simt.com/cgi-bin/p/awtp-custom.cgi?d=simt&page=30279

Virtual Reality Society (n.d.). History of virtual reality. Retrieved from https://www.vrs.org.uk/virtual-reality/history.html

Warburton, S. (2009). Second Life in higher education: Assessing the potential for and the barriers to deploying virtual worlds in learning and teaching. *British Journal of Educational Technology, 40*(3), 414–426. doi:10.1111/j.1467-8535.2009.00952.x

Epilogue

Virtual reality (VR) has been around for decades in various forms. However, today VR is beginning to make an impact in higher education. In the future, 2020 may become the year referred to when this extremely powerful teaching method made its debut. As can been understood from the chapters in this book, VR is already found in community colleges and universities throughout the nation. It is developing behind the scenes with no great fanfare at the moment, but its potential is enormous and powerful. It has the potential to become the major method of learning for years to come, at least until scientists figure out how to put everything one needs to know into the mind directly.

Remember the development of Uber, Airbnb, and Amazon in the early part of the 21st century. Each began as an idea and each developed in the background for a time in a stage 1, the developmental stage. During this time, these businesses were not particularly noticed or of great concern. When each of them moved to stage 2, the exponential stage, they grew phenomenally rapidly and extensively. Uber, Airbnb, and Amazon went international in a matter of a few years and changed things dramatically.

Virtual reality in education is on that same path. At the moment, VR is in stage 1, development, and moving toward stage 2, exponential. When it hits the exponential stage, it has the potential to change education completely, and for those who were not paying attention, it will seem like it came from nowhere. Educators at all levels need to be preparing now for stage 2 of virtual reality by learning about it, watching its development, and experimenting with its possibilities, so that they are not taken by surprise when it arrives. It is coming as sure as the sun rises and the New Year approaches. Do not be caught unaware.

About the Editor and Contributors

Darrel Staat was born in Chicago, Illinois, and grew up in Holland, Michigan. He received his bachelor's degree from Hope College in his home town, his master's degree from Western Michigan University in Kalamazoo, and his doctorate from the University of Michigan in Ann Arbor. He spent a number of years as a teacher in public schools and community colleges before he ventured into community college administration.

He served as president of Eastern Maine Community College in Bangor, Maine, for three years. He was the founding president of York County Community College, in Wells, Maine, for 5 years. He served as president of Central Virginia Community College in Lynchburg, Virginia, for 11 years. He served as president of the South Carolina Technical College System for 3 years and then retired.

Not finding retirement to his liking, he assumed the position of coordinator and professor of the Higher Education Executive Leadership Program at Wingate University in Wingate, North Carolina, where he is now in his seventh year. He has published five books on the topic of technology and higher education in the 21st century. These include:

- *Higher Education Planning in an Exponential Age: A Continuous, Dynamic Process* (2021)
- *Student-Focused Learning: Higher Education in an Exponential Digital Era* (2020)
- *A Baseline of Development: Higher Education and Technology* (2019)
- *Exponential Technologies: Higher Education in an Era of Serial Disruptions* (2019)
- *Facing an Exponential Future: Technology and the Community College* (2018)

***

Kenneth N. Buckey holds a master's degree in management and an EdS in higher education administration. He is currently the department chair of the Fire Protection and Emergency Management Department at Fayetteville Technical and Community College in Fayetteville, North Carolina.

Candice B. Lewis holds a master's degree in radiologic sciences and an EdS in higher education administration. She is currently the dean of the School of Health Sciences at Greenville Technical College in Greenville, South Carolina.

Lavar N. Marlowe holds a master's degree in Christian education and an EdS in higher education administration. He is currently an instructor at Brunswick Community College in Bolivia, North Carolina.

Jeffrey Merritt holds a master's degree in business administration and an EdS in higher education administration. He is currently the dean of Educational Services and Technology at Wake Technical and Community College in Raleigh, North Carolina.

Laneika K. Musalini holds a master's degree in human resources development and an EdS in higher education administration. She is currently the director of grants development at Tri-County Technical College in Pendleton, South Carolina.

Chris R. Nesbitt holds a master's degree in business administration and an EdS in higher education administration. He is currently the director of law enforcement training and chair of the Criminal Justice Technology Department at Rowan-Cabarrus Community College in Salisbury, North Carolina.

Shelly Stone-Moye holds a master's degree in social work and an EdS in higher education administration. She is currently the vice president of student development at Piedmont Community College in Roxboro, North Carolina.

Tou Vang holds a master's degree in public health in community health education and an EdS in higher education administration. He is currently the admissions coach at Western Piedmont Community College in Morganton, North Carolina.

www.ingramcontent.com/pod-product-compliance
Ingram Content Group UK Ltd.
Pitfield, Milton Keynes, MK11 3LW, UK
UKHW012042150926
475560UK00001B/149

9 781475 8612